A TREMOR
IN THE BLOOD
USES AND ABUSES OF THE
LIE DETECTOR

A TREMOR
IN THE BLOOD
USES AND ABUSES OF THE
LIE DETECTOR

DAVID T. LYKKEN

PERSEUS BOOKS
Reading, Massachusetts

Library of Congress Cataloging-in-Publication Data

On file

ISBN 0-306-45782-2

Perseus Books is a member of the Perseus Books Group.

The first edition of this book was published by McGraw-Hill, Inc., New York, 1981.

Printed in the United States of America

The first edition of this book was for my sons:
Jesse, Joseph, and Matthew.

This new edition is for *their* sons:
Zeke, Jake, Ezra, Adin, Oliver, Erik, and Karl.

Guilt carries Fear always about with it; there is a Tremor in the Blood of a Thief, that, if attended to, would effectually discover him; and if charged as a suspicious Fellow, on the Suspicion only I would always feel his Pulse, and would recommend it to practice. It is true some are so hardened in Crime that they will boldly hold their Faces to it, carry it off with an Air of Contempt, and outface even a Pursuer; but take hold of his Wrist and feel his Pulse, there you will find his Guilt; . . . a fluttering Heart, an unequal Pulse, a sudden Palpitation shall evidently confess he is the Man, in spite of a bold Countenance or a false Tongue.

—DANIEL DEFOE, *An Effectual Scheme for the Immediate Preventaion of Street Robberies and Surpressing All other Disorders of the Night,* 1730

CONTENTS

PREFACE xv

Chapter 1   PROLOGUE 1

Part I   LIE DETECTION: THE CONCEPT 5

Chapter 2   MR. READER HAS A CHANCE TO
             PROVE HIS INNOCENCE 7

Chapter 3   A BRIEF HISTORY OF LIE DETECTION 21
             William Moulton Marston 25
             Larson and the Berkeley Developments 27
             The Chicago Period 29
             Reid's Contribution: The Clinical Lie Test 30
             Backster and the Control Question Test (CQT) ... 32
             The Boom and Bust of the Polygraph Industry ... 34
             The Guilty Knowledge Test (GKT) 38
             Voice Analysis 41
             Honesty Tests 43
             The Tools of Diogenes 44
             Science and the Lie Detector 49

Chapter 4   THE TRUTH VERIFIER 53
               The Dawning of the Age of Truth 54
               Is There a Truth Verifier? 60
               The Current Status of the
                 Specific Lie Response 63

Chapter 5   EVALUATING THE EVIDENCE 67
               The Limitations of Expert Opinion 68
               How Polygraph-Induced Confessions
                 Mislead Polygraphers 70
               Lykken's Law 74
               Reliability versus Validity 76
               Assessing Lie Detector Validity 84
               Determining Ground Truth 85
               Summary 88

Part II   LIE DETECTION: THE METHODS 89

Chapter 6   THE CLINICAL LIE TEST: THE EXAMINER
                 AS "LIE DETECTOR" 93
               Subjective Scoring and "Behavior Symptoms" 95
               Assumptions of the Clinical
                 Polygraph Examination 98
               Validity of the Clinical Lie Test 104
               Verdict 106

Chapter 7   THE RELEVANT/IRRELEVANT (R/I) TEST 109
               Assumptions of the R/I Test 110
               Validity of the R/I Test 112
               Verdict 113

Chapter 8   THE CONTROL QUESTION TEST (CQT) 115
               A Genuine Control Question Test 117
               Assumptions of the Control Question Test 119
               Some Real-Life Examples 124

The Validity of the Control Question Test 128
An Example of a Bad Validity Study 130
CQT Studies Published in Scientific Journals 132
Verdict 135

Chapter 9   THREE DIFFERENT VERSIONS OF THE CQT . . . 137
The Directed Lie Test (DLT) 137
The Positive Control Test (PCT) 140
The Truth Control Test (TCT) 143
The Peak of Tension Test (POT) 147

Chapter 10   POLYGRAPH SCREENING TECHNIQUES 151
The Format of the Screening Test 156
Assumptions of the Polygraph Screening Test 158
The Validity of the Polygraph Screening Test 160
Verdict 161

Chapter 11   VOICE STRESS ANALYSIS 163
A Tremor in the Voice: The PSE 165
Does the PSE Measure Stress? 166
Does the PSE Detect Lying? 167
Recent Developments 171
Verdict 172

Chapter 12   SURVEYS OF SCIENTIFIC OPINION
               OF THE "LIE DETECTOR" 175
Two Prior Polls 177
Two Recent Polls 179
Information Provided to Respondents 179
How the Surveys Were Conducted 182
Results 182
Summary 186

Chapter 13   THE TOOLS OF DIOGENES: AN OVERVIEW . . . 189
Lies, Lies, Lies! 191

Part III   **LIE DETECTION: THE APPLICATIONS** **195**

Chapter 14   TRUTH, LTD.: THE POLYGRAPHER
              AS PRIVATE DETECTIVE   197
              Intramural Crime   198
              The Coker Case   200
              A Question of Rights   207
              To Catch a Thief   210

Chapter 15   PREEMPLOYMENT SCREENING BY THE FBI
              AND OTHER FEDERAL AGENCIES   213
              The Case of Major C   214
              Cry Havoc   216
              Screening FBI Agents   218

Chapter 16   HONESTY TESTING: AN ENVIRONMENTAL
              IMPACT ASSESSMENT   223
              The Problem   226
              The Integrity Screening Approach   227
              Beating the "Honesty" Tests   230
              The Good Management Approach   232
              Summary   233

Chapter 17   THE FOURTH DEGREE: POLYGRAPHICALLY
              INDUCED CONFESSIONS   235
              The Peter Reilly Case   240
              Why Do People Confess?   242
              Confessions and the Courts   243
              The Embassy Marine Guard Scandal   245

Chapter 18   THE LIE DETECTOR AND THE COURTS   249
              The Frye Case   250
              Allegations of Sexual Abuse   251
              The Polygraph for the Defense   253
              The Question of Base Rates   256
              The "Friendly Polygrapher"   259

The Polygrapher as Expert Witness 260
How Juries React to Lie Test Evidence 262
The Psychopathic Liar 267
The Polygraph in Criminal Investigation 271

Chapter 19   HOW TO BEAT THE LIE DETECTOR 273
Methods of Beating the Lie Detector 274
Methods Taught by Floyd Fay 275
Methods Taught by the Raskin Group 276
The Right Way to Beat the Polygraph 277

Part IV   DETECTING GUILTY KNOWLEDGE 281

Chapter 20   ORIGINS OF THE GUILTY
KNOWLEDGE TEST 283
The Experiment 283
The Results 286
Replication by Other Investigators 287
A Good GKT 288
Beating the GKT 292

Chapter 21   FORENSIC USES OF THE GUILTY
KNOWLEDGE TEST 295
Did O. J. Simpson Kill His Wife? 297
A Guilty Knowledge Test for O. J. Simpson 298
Who Blew Up the Murrah Building? 299
Scoring the GKT 301
Countermeasures 303
The GKT and the Polygraph 303
Forensic Applications of the GKT 304
The GKT in the Courtroom 306
Future Prospects 307

NOTES AND REFERENCES 309

INDEX 327

PREFACE

The first edition of this book was also the first and only monograph on this important topic written by a scientist for the edification of interested non-scientists, including lawyers and policy-makers. When Michael Hennelly of Plenum suggested that I consider preparing a new edition, I realized that many interesting developments had occurred in this field in the past two decades. No actual "lie detectors" have been invented but some of the old ones have been patched up, relabeled, and touted as improvements. The myth of the polygraph has been married to the mystique of the computer, a dangerous liaison that has spawned a litter of mischievous mythlets. Since 1980, I have testified as an expert witness for the purpose of impeaching lie detector findings in criminal trials and courts-martial from Alaska to Florida. I have been contacted by hundreds of victims of mistaken lie detector tests or by their attorneys. Many of their stories provide illustrative material—illustrations that I think are truly disturbing.

I have testified in support of anti-polygraph legislation before committees of state legislatures, the United States House and Senate and I hope I contributed in a small way to the eventual passage of the federal Employee Polygraph Protection Act of 1988. This statute prevents most employers in the private sector from requiring job applicants or current employees to submit to polygraph screening tests. Police and all federal employees are excluded from this protection, however. The result has been that many honorable people, the people I would like to see on our police forces and in our federal police and security agencies, have been falsely branded and excluded as "deceptive" by arrogant polygraph examiners. Polygraph examiners, most of whom are honorable people, are made arrogant by the fact that they are protected from discovering their mistakes. The only feedback examiners typically receive is when a subject whom they have accused of lying actually confesses and thus corroborates their diagnosis. As shown in this new edition, these occasional verifications are entirely compatible with the possibility that the lie detector is no more accurate than the toss of a coin.

Moreover, under the 1988 statute, it is still permissible for employers complaining of losses or sabotage to call in—not the police—but private security people who may require suspected employees to take lie detector tests. Acting as both judge and jury, the polygraph examiner then identifies an alleged culprit who can be dismissed or punished without benefit of due process. For reasons explained herein, we have no definitive scientific evidence on which to base precise estimates of the lie detector's validity. But we have enough evidence to say that an innocent person has nearly a 50/50 chance of failing a lie detector test, odds that are much worse than in Russian Roulette.

A different method of polygraphic interrogation, designed to detect guilty knowledge rather than lying, was described in the first edition but I can now report encouraging new data concerning its validity. I can also report that, of scientists who are knowledgeable about the theory and methods of polygraphy, the vast majority agree with the views and conclusions expressed here. Thanks to my former student and now colleague, W. G. Iacono, Distinguished McKnight Professor of Psychology, and to his student, Deborah Rasmussen, I am able to present here the results of polls detailing the opinions of members of the Society for Psychophysiological Research and of psychologists distinguished by election as Fellows of Division 1 (General Psychology) of the American Psychological Association.

However, I should hasten to explain that the theory and methods of polygraphic lie detection are not rocket science, indeed, they are not science at all. Most of these techniques were developed by police or by lawyers and they involve assumptions so implausible that any lay person (including most jurors) can understand and dismiss them with very little expert help. It is my hope that this new edition will provide hope to the victims and an incentive to citizens, courts, and legislators finally to relegate the "lie detector" to some dusty shelf in the basement of the Smithsonian Institution.

Chapter 1

PROLOGUE

*It is a pleasure to stand upon the shore, and to see ships tost
upon the sea; a pleasure to stand in the windows of a castle, and
to see a battle and the adventures thereof below: but no pleasure
is comparable to the standing upon the vantage ground of truth
(a hill not to be commanded, and where the air is always clear
and serene), and to see the errors, and wanderings, and mists,
and tempests, in the vale below.*

—Francis Bacon

*One of the most striking differences between a cat and a lie is
that a cat has only nine lives.*

—Mark Twain

Thousands of Americans are subjected (or subject themselves) to polygraph or "lie detector" tests each year. When the first edition of this work was published in 1981, that estimate was in the *hundreds* of thousands, but due to the Employee Polygraph Protection Act that was passed by Congress and signed by President Reagan in 1988, it is now illegal for most employers in the private sector to require employees or job applicants to submit to this intrusive and humiliating procedure. I like to think my book contributed to the change. Unhappily, however, the 1988 statute exempts law enforcement and other government employees from its protection, and as we see in Chapter 13, employers in the private sector still can call in the "polygraph police" under certain circumstances.

1

Your turn to be strapped into the lie detector apparatus may come in any one of several ways. You may become a suspect in a criminal investigation—many innocent people just like you have never expected to find themselves in such a situation. Nurse Francine Bronson of Yakima, Washington, never imagined in her wildest dreams that her ex-husband's new wife would tell the police that she thought Francine had been sexually abusing her own four-year-old son. Floyd Fay, a construction worker in Ohio, did not expect to find police officers on his doorstep at 4:30 A.M., guns drawn, ready to arrest him for the murder of an acquaintance.

If you are a woman, you may have the great misfortune to be sexually attacked by someone you know—"date rape"—and then discover that the district attorney in your jurisdiction will not pursue your complaint unless you submit to (and pass) a lie detector test. Winnefred R. of Charleston, South Carolina, did not recognize the masked man who entered her house one night, forced her to take a soapy bath, and then carefully shaved off her pubic hair before raping her. But she recognized his distinctive, lisping voice a few days later when he entered the store where she worked. The local sheriff refused to investigate her complaint until she had submitted to a lie detector test.

You may become embroiled in civil litigation, either as plaintiff or defendant, in which your version of events conflicts with that of the other party. Your attorney or your opponent's counsel may suggest that you both take lie detector tests with each side agreeing in advance that the results will be admissible in court. Perhaps your spouse will come to suspect you of infidelity and one of you will conclude that the only way to clear the air is by lie detector.

You may apply for a job in law enforcement and discover that the last step in the application procedure is a polygraph test. Elizabeth M., of San Francisco, had aspired from girlhood to work for the FBI. After graduating from law school, she applied to the bureau, sailed through the initial assessment procedures, then reported as directed for her polygraph test. Because the examiner concluded she was deceptive in denying any use of illegal drugs, Elizabeth's career aspirations evaporated on the spot. This was ironic because Elizabeth, whose policeman father had told her many stories of the ravages of narcotics he witnessed professionally, had a genuine phobia of street drugs.

If you work for a federal security agency such as the CIA, the National Security Administration, the Drug Enforcement Administration, or the Secret Service, and have managed to pass the initial polygraph screening required by these agencies, you may be required to take periodic polygraph examinations. Aldrich Ames, who sold secrets to the Soviets for years while employed by the CIA, driving to work each day from his elegant home in his expensive car, succeeded in his spy career for as long as he did because his ability to beat the lie detector deflected official suspicions. Major C, a West Point graduate and Soviet area specialist who was serving as a translator on the Washington–Moscow hotline during the Reagan administration, deserved his Top Secret security clearance because he was a straight arrow and a patriot. Yet, on a routine polygraph screening test, the examiner concluded that Major C's "No" answer to the question "Have you discussed any secret information with a foreign national?" was a lie. I first heard the outline of this major's story when he called me to ask about the chapter in my book that explains how the lie detector can be beaten. In his next call, he reported that "The only hard part was keeping a straight face."

Many people think, mistakenly, that the results of lie detector tests are not admissible as evidence in U.S. courts. Although the supreme courts of some more enlightened states have ruled polygraph tests inadmissible in criminal trials,[1] they can be introduced into evidence in about half our states under certain circumstances, and in American military courts-martial, lie test results are frequently admitted. Ominously, a 1993 U.S. Supreme Court decision,[2] although it did not address polygraph tests specifically, now obliges federal courts, and state courts that follow the federal lead, to consider on a case-by-case basis all offers of polygraph evidence by either the prosecution or the defense. Across the country, prosecutors, both state and federal, have learned that when their case is too weak to go to trial, it may be a useful strategy to offer the defendant what sounds like a reasonable proposition: "If you are willing to demonstrate your innocence by taking a polygraph test, then, if you pass, we will drop the charges. However, you must agree in advance that the polygraph results can be used in evidence against you should you be found to be deceptive." Innocent suspects who, like most Americans, believe in the myth of the lie detector, are inclined to accept such offers, often to their subsequent dismay.

There are many psychological tests in use today that play important roles in people's lives—aptitude tests used in selecting employees, intelligence and achievement tests that play a role in determining admission to colleges and professional schools, personality and mental function tests used for diagnostic purposes. Although the polygraph test employs measures of physiological reactions, it too is a psychological test that purports to assess the examinee's state of mind. And the outcome of a lie detector test can determine whether you get or keep your job, whether you will be prosecuted for a crime, or, if prosecuted, whether you will be convicted. Judged on this basis, polygraph testing is an important example of applied psychology. It may seem surprising, then, that the lie detection methods in wide use today were invented by lawyers and policemen, not by psychologists, and that only a handful of practicing polygraphers have any significant psychological training. It is more surprising still—and, indeed, it is a cause for considerable alarm—that after more than 70 years of use, there is as yet no scientifically acceptable estimate of the accuracy of polygraph test results.

There have been a few good studies, however, that permit us to make upper-limit estimates of polygraph accuracy, and, moreover, the assumptions of the various techniques are simple enough so that both laypersons and psychological scientists can make intelligent evaluations of their plausibility. We shall see later that the assumptions of the various techniques of lie detection now in use are plainly unrealistic and lacking in plausibility. We shall also see, however, that there is another type of polygraph test—designed to detect, not lying, but rather the possession of guilty knowledge—that is based on reasonable assumptions and that shows real promise as a useful tool of forensic science.

Part I

LIE DETECTION: THE CONCEPT

He who would distinguish the false from the true
Must have an adequate idea of what is false and true.

—SPINOZA, *Ethics*, 1677

Whenever I lecture on polygraphic interrogation, I like to begin by asking the audience to indicate by show of hands how many would agree to take a lie detector test—and how many would refuse—in some plausible, hypothetical situation that I briefly describe. Among American audiences, a substantial majority acknowledge that they would take the test if they were innocent of wrongdoing and wished to prove their truthfulness. Many Europeans, in contrast, have never heard of any such device. The lie detector is almost exclusively an American artifact, and in the United States, it seems firmly entrenched in popular mythology. There lurks some vague familiarity with the concept in the mind of nearly every American who can read and wears shoes. The history of this phenomenon, which has a kind of fascination, is briefly reviewed in Chapter 3. That chapter also previews the armamentarium available to the modern Diogenes.

But before poisoning your mind with facts and figures, I have arranged to let you experience a lie detector test yourself. Chapter 2 dramatizes what might ensue if you were asked to take a lie detector test administered by a highly competent, professional police examiner under real-life conditions. Chapter 4 considers whether there is now or will ever

5

likely be a literal Truth Verifier, a machine that can detect lying. The answer, quite certainly, is in the negative. There is no lie detector *machine* but only a form of interview or test employing a machine known as a polygraph, which provides certain data that the examiner uses in reaching his diagnosis of truthful or deceptive. But there are several different types of lie detector tests, and Chapter 5 discusses the problem of assessing the accuracy of these tests in general: what kind of evidence is relevant to what, and what "evidence" is not relevant at all. Chapter 5 also contains some discussion of elementary statistical concepts. Certain researchers in this field have confused the ideas of "reliability" and "validity" and have interpreted the numbers from the research findings incorrectly, so it seems to me essential to set forth the appropriate methods of interpreting the data and to show why they are correct. The ideas involved are not difficult and I have tried to present them in such a way that even the number-phobic reader can follow the argument.

Chapter 2

MR. READER HAS
A CHANCE TO PROVE
HIS INNOCENCE

*We honor Washington because he could not tell a lie. Mine is a
much harder case; I can, but I won't.*

—Mark Twain

*And it must follow, as the night of day, thou canst not then be
false to any man.*

—Shakespeare, *Hamlet*

Later on in this book I consider polygraphic interrogation from the point
of view of the examiner, who wants to produce accurate results, and from
the perspective of employers, the police, and the courts, who wish to make
decisions based on those results. I also examine the lie detector industry
from the viewpoint of the citizen and his elected representatives, who
must be concerned with the impact on American society of the present
rapid growth and outreach of this technology. But first, dear Reader, you
should try to get a sense of what a lie detector test is like for the person
being tested, whom I refer to variously as the subject or as the respondent.
Let us set up a scenario in which you are the subject and I am the
polygraph examiner. My files are replete with real-life examples we might

use; I describe a number of these cases in subsequent chapters. But most of these cases have unhappy or, at best, ambiguous outcomes, so their use would paint an unfair picture of the way expert polygraph examiners conduct their business. Therefore, I shall invent a situation for our present purposes.

Let us assume that you are an adult male employed in some white-collar position. Something of value disappears from your place of business; let us say an envelope containing $2,000 in cash is missing from the office safe. There are no signs of breaking and entering. The police consider it to be an "inside job" and suspicion centers on the seven employees who have access to the safe and know the combination. You are one of these seven. Your employer calls you in and explains the situation. The police are asking each of the potential suspects to agree to take a polygraph examination. You cannot be forced to take the test, but unless you have a contract or belong to a strong union, you might lose your job if you refuse. It is assumed that if you are innocent, you will welcome this opportunity to prove it. An appointment is made for you to visit my office at police headquarters at 10 A.M. the next day.

Next morning, in the waiting room, you find among the reading matter books and pamphlets bearing such titles as *Polygraph for the Defense* or *The Polygraph Story: Dedicated to Man's Right to Verify the Truth.* My diplomas and credentials line the walls. After some minutes, I come in and introduce myself. I am conservatively dressed and my manner is polite and professional as I usher you into the adjoining polygraph room. This room is carpeted and quiet with a few pastoral prints on the walls. Toward one end you notice an ordinary desk bearing a complicated-looking instrument in an aluminum case, the size of a small suitcase. This is the polygraph. Centered against the front or closed side of the desk is a padded vinyl armchair facing toward the open room. If you are seated there, the desk and polygraph are out of direct view. I pull up a chair so that we can talk comfortably.

"*Now then, your full name is Robert S. Reader and you live at 123 Sunflower Lane?*" I have a clipboard on my lap and make notes as we go.

"*What was your birthdate, Mr. Reader? Are you married? How long have you been married? Any children? And what are their names and ages?*" Routine questions, breaking the ice; you're beginning to relax a little.

"*How long have you been with the Frobischer Company? I see, and you're now the assistant office manager, is that correct? Now, on the day that the money was discovered missing, that would be March 19th, Tuesday, did you come to work*

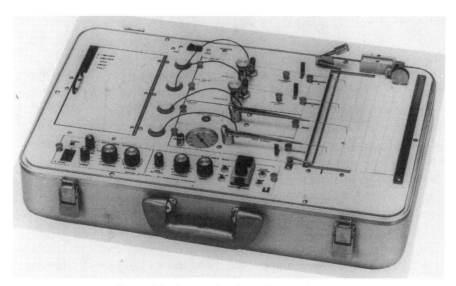

Figure 2.1. A typical polygraph instrument.

as usual? And did you have occasion to open the safe or to take something out or put something in? You did not? Okay. Have you any recollection at all of anything unusual involving that safe during that morning? For example, did you notice that anyone had stepped away from it for a time, leaving it open? I see. So then you have no reason for thinking that someone who didn't know how to unlock the safe might have been able to get into it that morning?

"*All right, Mr. Reader. Now I realize that you don't want to get any of your colleagues in trouble but the fact is that this test you're going to be taking might be influenced by any suspicions you have about who might have taken the money or where it is now, even if your doubts don't have any real basis. So I want you to tell me now about any thoughts along these lines you might have so you can get them off your chest before we do the test. Who do you think took that money? You're sure you have no idea at all, no suspicions for any reason?*

"*Okay. Now, I have a list of questions here that I plan to ask you when we've got you hooked up to the polygraph, and I want to go over the list with you first to make sure you understand each question and that you can answer each of them with a simple yes or no. The first one is an irrelevant question that we ask just for calibration purposes. I'll ask you, 'Is today Tuesday?' and you will answer 'Yes.' Then I'll ask you, 'Are you afraid I might ask you about something other than the questions we have already reviewed?' and I want you to be able to answer that one*

'No,' okay? Next, I'll ask you, 'Do you know for sure who took the money from the safe?' From what you've told me, I gather you can answer that one 'No' also with a clear conscience, right?

"All right, Mr. Reader, now whoever took this money, it's unlikely that it was the first really dishonest thing that person ever did. Therefore, during the test, I'm going to ask you some questions to find out what kind of a person you are, whether in the past you have ever done anything that might indicate you're the sort of person who would steal money from the office safe. For example, take your income tax returns. I suppose most people make mistakes on their income tax now and then, forget to put something down or exaggerate their expenses. But you haven't been deliberately cheating on your tax returns, have you? If I ask you, 'Over the past five years, have you ever deliberately cheated the government out of more than $100 on your income tax?' you could answer that 'No,' couldn't you?"

Now this might make you begin to feel a little uncomfortable. You're no tax criminal and when you were called in for an IRS audit four years ago you never felt an impulse to skip out of the country. But on the other hand, when you win a few dollars on a football pool or in a poker game, it has never occurred to you to put it down as taxable income. And when you estimate mileage or business expenses, you don't tend to choose the most conservative figures. But you don't want this straight-looking man with his neat haircut to get the wrong idea about you.

"Yes, sure, I could answer that 'No.'"

"That's fine. Then I will ask you, 'Did you take the money from the office safe?' and, of course, you'll answer 'No,' right? Let's be sure there is no ambiguity about that question; I'm talking about the $2,000 in cash that was in the safe in an envelope and which disappeared on March 19th, okay?

"Now, Mr. Reader, as you may have heard, there was another theft in your office last year that was never solved. Mr. Hodsdon had one of those fancy wristwatches with a solid gold bracelet that he left either in the washroom or on his desk and when he looked for it, it was gone, and he never got it back. He didn't do anything about it at the time, but he's asked us to inquire about that, too, while we're giving these tests. So, I'm going to ask you, 'Did you take Mr. Hodsdon's gold watch?' Are you comfortable with that question?"

This is the first you've heard anything about a missing watch. There really must be a thief in the office. Funny that old Hodsdon didn't make a fuss about it at the time; he raises hell if his newspaper is missing.

"Sure," you say. "I never even knew about the watch."

"Now we'll have another general honesty question, Mr. Reader. Suppose I

ask you, 'From the time you graduated from high school up to two years ago, did you ever steal anything of value?' Would you be able to answer 'No' to that?"

"Oh, well," you protest, "that's pretty tough. Everybody probably takes some little thing that doesn't belong to him once in a while."

I give you a rather searching look. *"You say 'some little thing.' Maybe we could change the question to leave the 'little things' out of it. Suppose I say, 'Between high school and two years ago, have you ever stolen anything worth more than $10?' Could you answer 'No' to that?"*

You aren't a thief and there is nothing in particular burning in your mind at the moment, but you'd hate to stake your life that you won't start thinking of things by this afternoon. "Look," you say, "I can't think of anything right now worth even $10, but you're talking about quite a few years there."

"Mr. Reader, I'm just trying to get a wording that you're comfortable with. It's just a matter of choosing a figure. If you've previously stolen something worth hundreds of dollars, then, of course, we would have to consider that such a person might also steal a watch or some money from the safe. I assume you haven't done that but then you say that $10 might be too low. What sort of figure would you like to have me use in this question?"

You try to remember the dividing line between petty larceny and grand theft. "How about $25? I think I could be comfortable with $25." You walked in here feeling like an assistant office manager on his way up; now you're beginning to feel rather like a petty crook.

"All right, that's the way we'll do it. 'Between high school and two years ago, have you ever stolen anything worth more than $25?' And you will answer 'No' to that. Then I will ask you, 'Regarding that money that was taken from the office safe, did you take that money?'—and, again, you will answer 'No.' Then we have to have one more general honesty question. I'll ask you, 'Between high school and two years ago, did you ever tell a lie to someone in authority in order to stay out of trouble?' What I'm obviously getting at here is whether, when you do make a mistake, you own up to it or whether you're the kind of person who will try to lie his way out of it. Will you be able to answer 'No' to this question?"

"Yes, certainly, no problem about that." Again you have the feeling that some awful contradicting recollection is lurking in the back of your mind, but this man is probably already wondering about you and you don't want to give him a worse impression than you have already.

"Okay, Mr. Reader, that makes nine questions altogether. Now we'll go through this list several times while you're hooked up to the polygraph. I'll pause

15 or 20 seconds between questions and when we get to the end of the list, I'll turn the machine off and you can have a little rest before the next test. I may change the order of the questions, but I won't ask you anything we haven't already agreed on. Or, if I do want to add some new questions, I'll turn off the machine and go over them with you first and make sure that you're agreeable before I use them. Also, when I first hook you up, I'm going to go through a special calibration test in which I'll have you choose a card from a deck and then I'll try to find out which card you've chosen by asking you questions and looking at your polygrams.

"Now, let me explain the polygraph to you. There are four attachments to your body so that the machine can pick up your emotional reactions to the questions. There are these two soft rubber belts that I'll strap around you, one around your stomach and one around your chest. They will measure your breathing. Then there are these two wires with these little attachments on the end. I'll fasten them on the ends of two of your fingers. They measure sweating responses. Finally, I'll wrap this blood pressure cuff around your upper arm here. While I'm running a test, I'll pump up the pressure in the cuff so it will feel tight around your arm and then I'll relax it again between tests. All this is perfectly safe, of course, and none of it will hurt at all."

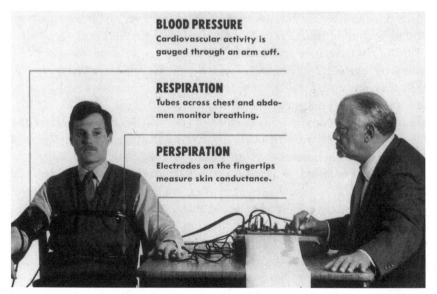

Figure 2.2. From *Discover* magazine, © 1986 Kay Chernush, reprinted with permission.

Now I hand you a printed consent form and explain that you must read and sign it before we can proceed. Most consent forms used by polygraphers are remarkably generous in waiving the rights of the person to be tested while at the same time being rather vague about the responsibilities of the examiner. Since we are trying to run this imaginary test in a professional manner, let us assume that my consent form has been designed with your, the respondent's, rights uppermost in mind.

CONSENT FORM

I hereby consent to a polygraph examination to be administered by (name of examiner) on this date of (date) and to having the necessary polygraph attachments made to my body as has already been explained to me. I understand that the examination is to be concerned with (subject of examination) and that the only questions that will be asked me during any portion of the examination will be discussed with me and approved by me prior to the test. I understand that an audiotape recording is being made of the entire test, including my discussions with the examiner before and after the actual testing, and that I may obtain a copy of that recording, at my own expense, at any time within six months from the present date. I hereby authorize the examiner to communicate the results of the examination to the police officer in charge of the investigation of the above-named matter and also to (enter name or "no other person").

(Signature of respondent)

"This 'no other person,' " you ask, "does this mean that you won't even tell my boss how I did on this test?"

"That's right, Mr. Reader. This test is for police purposes and police use only." I do not explain it in detail but the fact is that this is an important protection for you. Suppose you fail this test while your six colleagues pass theirs. And suppose you continue to protest your innocence and the police are unable to find any evidence against you. You cannot be convicted, indicted, or even arrested just on the basis of a lie detector test, not in this jurisdiction. But if the results are given to your employer, you will very probably be fired. You will be punished, your career perhaps permanently blighted, with no vestige of due process. If I were a private examiner, hired by Frobischer, Inc., then you would have to take your chances. But (in this scenario) I am a police examiner and my job is to assist a criminal investigation in an orderly and legal fashion.

Now I apply the polygraph attachments and you begin to feel uncomfortable again, exposed in a curious way, those four pens behind you scratching out their squiggly lines that presumably reveal what's going on inside you, your emotional reactions, almost reading your mind. It is years since you've felt as self-conscious as you feel right now. You think of a time when your fourth-grade teacher accused you of some misdemeanor that you hadn't committed and, although innocent, you could feel yourself blushing under her hard gaze and you knew you looked guilty. You begin to wonder whether this machine can tell an honest emotion from a guilty one.

I take a deck of playing cards out of the desk, shuffle them, and fan them out in front of you face down. *"Choose a card, look at it, and then put it in your pocket without letting me see it."*

Your card is the queen of hearts. You tuck it away in your shirt pocket.

"All right, Mr. Reader, I'm going to ask you some questions about the card you have in your pocket. I want you to answer 'No' to each question no matter what the true answer might be. Put both feet on the floor and sit there quietly until I tell you you can move. I'll pump up the pressure on the arm cuff now. Remember to answer 'No' to each question. Here we go."

"About the card in your pocket, is it a red suit?"

"No," you answer.

"Well, that looks like a deceptive response," I say after a few seconds of watching the polygraph behind you. *"I'm going to assume that it is a red suit. Is the card in your pocket a diamond?"*

"No."

Another pause. *"Is it a heart?"*

"No."

"All right, I think the suit must be hearts. Is the card in your pocket a face card?"

"No."

Ten or fifteen seconds elapse after each answer. The cuff on your right arm seems tight. Your nose itches and you would like to shift your weight in the chair but you are not supposed to move.

"Well, I think this is going to be an easy one. You're a good responder, Mr. Reader. Your card is a face card all right. Is the card in your pocket the king of hearts?"

"No."

Another pause. *"Is it the queen of hearts?"*

"No." You have a feeling that the jig is up.

"Okay, Mr. Reader, I'll release the pressure on your arm now and you can move in your chair if you want to. We're done with this calibration test. Your card is the queen of hearts—right?" You hand it over to me. *"Now I know what your*

chart looks like when you're telling the truth and what it looks like when you're lying. So now in just a minute or two we'll get started with the first test."

This little demonstration has been quite effective. You still feel a bit naked, connected to that machine, but it does seem to work. Nevertheless, you'll be glad when this is over. I pump up the arm cuff again and proceed to work through the question list, just as we had discussed it earlier. This takes about four minutes. It is a relief when I release the cuff pressure again and you can flex your arm and restore circulation. During this rest period, I stay behind you looking at the output of the machine and I don't say much.

"We got a good record that time, Mr. Reader. You're sitting nice and quietly and you give good responses. Settle down now and we'll go through the questions again."

This time I change the order of some of the questions but there aren't any new ones. The two questions about the money from the safe and the one about old Hodsdon's watch continue to alternate with what I called the "general honesty" questions. I give you another two-or-three-minute rest after this set while I stand behind you at the polygraph making notes on the charts.

"Now for this next test, I want you to listen carefully to each question but I don't want you to answer. Don't say anything at all. Just sit there quietly and listen to the questions. Here we go."

Let us get out of the scenario for a moment while you consider how you think you might be feeling at this stage of the polygraph test. The nine questions are collected together for easy reference in Table 2.1. Those polygraph charts that I am studying behind your back will probably show some sort of reaction to each of the questions. What I am looking for is differences in the strength of reaction from one question to another. What would you predict I'll find? The first question, "Is today Tuesday?" is innocuous enough although you might show some small response since, being first, that question signals the start of the series. How might you have reacted to the questions about the money stolen from the safe? These, after all, are the reasons you are taking this test in the first place. On the one hand, the money was stolen, you can open the safe, you are a legitimate suspect. On the other hand, you are innocent; you did not take the money and you do not know who did. There is no logical reason for you to be upset by—to react emotionally to—these questions, is there? Remember that fourth-grade teacher looking down at you accusingly; you hadn't thrown the eraser but you blushed anyway. The polygraph is registering internal blushes; what will it find in you?

Table 2.1. Questions Used on Your Polygraph Examination

1. Is today Tuesday?
2. Are you afraid I might ask you about something other than the questions we have already reviewed?
3. Do you know for sure who took the money from the safe?
4. Over the past five years, have you ever deliberately cheated the government out of more than $100 on your income tax?
5. Did you take the money from the office safe?
6. Did you take Mr. Hodsdon's gold watch?
7. Between high school and two years ago, have you ever stolen anything worth more than $25?
8. Regarding that money that was taken from the office safe, did you take that money?
9. Between high school and two years ago, did you ever tell a lie to someone in authority in order to stay out of trouble?

Then there is Hodsdon's gold watch. You never even knew the thing was missing. What will the pens indicate when I ask you if you stole that watch? Now consider the "general honesty" questions, the ones about previous thefts, cheating on your income tax, lying to stay out of trouble. I told you I wanted to find out whether you were the kind of person who might be inclined to steal money from a safe and those questions bothered you some when we first talked about them. You weren't absolutely sure that your "No" answers were strictly true. What do you think the polygraph has had to say about your reactions to those questions as compared to the ones about the money in the safe or the watch?

What you would not have known, unless you had read this book or some other discussion of lie detection methods, is that I am going to evaluate your test by comparing your reactions to the relevant questions, the ones about the stolen money, with your reactions to the "general honesty" questions, which polygraphers call "control" questions. If you consistently respond more strongly to the relevant questions, then I will conclude that you are lying about the stolen money and you will fail the test; my diagnosis will be "Deception indicated." If you respond consistently more strongly to the control questions, I will report "No deception indicated." In the few cases where the two sets of reactions are about equal, I will score the test as "Inconclusive." How did you do? Did you pass? How will you do next time, now that you know how it works?

But let us return to that polygraph room, because the test isn't over yet. When I release the cuff pressure after the last question on the third, or "silent answer," test, there is quite a long silence while I stand there behind you, studying the charts.

"Look, Mr. Reader, on the matter of the gold watch, there doesn't seem to be any problem there. But on the questions about the money, I'm getting reactions to both of those. I'm wondering if there is anything about those two questions that might account for what I'm getting? Anything about the way they're worded? Maybe something about some previous experience with something from a safe or something stolen from the office or some currency you might have taken some other time? In other words, is there anything you haven't already told me about that might explain why you're reacting this way to those two questions?"

This is alarming news to say the least. No, you don't have any obsession about safes that you know of. You've never absconded with any handful of bank notes just as you didn't take this envelope of money from the office safe.

"No, I can't think of any reason. Except, after all, I do know that this money was stolen and that I'm a suspect, and I can't help feeling concerned about that."

"But what is there to be concerned about, as long as you didn't take the money?" I ask politely. *"I tell you what, Mr. Reader, I'm going to give you a few minutes to collect yourself and maybe you'll think of some reason why I'm getting these reactions. You just sit there quietly and think about it and I'll be back in a few minutes."*

I walk out of the room leaving you to stew. That large picture of a Swiss valley at the end of the room is actually a one-way mirror. In the darkened observation room on the other side of that wall I settle down to watch you squirm. Like most other polygraphers, I remember that marvelous case in Cincinnati from a few years ago.[1] A private examiner, testing an employee in connection with a theft of a few hundred dollars, had employed this same technique. The respondent, obviously upset, kept glancing back at the polygraph charts covered with their apparently damning hieroglyphics. Suddenly, as the examiner watched unbelievingly from the next room, the subject reached back, tore the charts from the machine, and proceeded to eat them, bit by bit, all six feet of paper six inches wide. After this bizarre meal was completed, the examiner returned as if nothing had happened and began to prepare for the next test. Suddenly he leaned his ear down to the polygraph and said,

"What's that? He ate them?!"

"My God, you mean the thing can talk, too?" expostulated the respondent, who then proceeded to confess that he had taken the money.

You don't do anything so dramatic. You just sit there, fidgeting from time to time. When I return, you have nothing to report, no explanation for those worrisome reactions to the critical questions. We do another test just like the first one. You are really nervous now; each time your heart beats you can see a slight movement of your shirt front and your voice sounds strange in your ears as you give your replies: "No," "No." When this fourth test is finished, I proceed to remove the polygraph attachments without saying anything. Then I pull over a chair in front of you, sit down, and look at you for a moment.

"Mr. Reader, we seldom have any criminals in this room. Just ordinary people who may have made a mistake, usually for a good reason. People who were pushed too far or tempted too much and then slipped up. And when ordinary decent people make a mistake, the thing for them to do—the only sensible thing to do—is to admit it, get it off their chests, make amends as best they can. Then they not only feel better about it, but the consequences of their mistakes will be minimized. Look at it from our point of view. A confession saves all kinds of time and hassle and we appreciate that. When somebody with a good record admits that he made a mistake, we appreciate that kind of cooperation, we think it shows a good attitude, and we try to arrange things in a way that makes minimum trouble for everybody. Now, have you anything you want to tell me about this matter? Anything at all?"

Let me cut off the narrative at this point before you say something you'll regret later. Let me make a confession instead. I was simply trying to give you an opportunity to confess, in case you had something to confess, and I might have done so no matter what your polygraph charts had looked like. As we see later, perhaps the major utility of the lie detector is its extraordinary capacity for eliciting confessions. Unfortunately, most polygraph examiners only interrogate in this way when the subject has reacted strongly to the "Did you do it?" questions and "failed" the test. This is because polygraphers genuinely believe in the technique, so when the response charts seem to them to indicate truthful answers, they see no point in interrogation. One interesting result of this practice is that the only time they get confessions is when the subject has "failed" the lie test, so each confession seems to validate their technique.

That story about Mr. Hodsdon's gold watch was pure invention, and it too is a little embellishment of mine that the professionals do not use. I wanted to see how you would react to a question about a crime you could not have committed since it never happened. If, on the other hand, you had reacted strongly to both the money and the gold watch questions, that would have suggested that you had *not* taken the money but were merely (and understandably) bothered by accusatory questions. On the other hand, if you had reacted strongly to the money question but not to the one about the watch, that might reasonably have made me think that perhaps you knew more about the missing money than you were letting on.

And one other thing. That deck of cards that we were using earlier? I bought it at a magic store. All 52 cards in the deck are queens of hearts.

Chapter 3

A BRIEF HISTORY
OF LIE DETECTION

Lord, Lord, how this world is given to lying.

—Falstaff in *Henry V, Part 1*

*All th' wurruld is busy deceivin' its neighbor an itsilf. Th' poor
are poor because they are poor liars an' th' rich are men that've
accumylated large stock iv non-assissable, inthrest-bearin' lies or
inherited th' same fr'm their indulgent an' mendacyous fathers.
That's what they tell me.*

—"Mr. Dooley" (Finley Peter Dunne)

*Some of the most amusing lies on record have been told in
connection with the Lie Detector itself.*

—W. M. Marston

Nature is full of guile. From the lowly insects to our primate cousins,
animals have evolved a wide variety of methods for deceiving other
animals, their enemies, their conspecifics and, most commonly, their prey.
Some of these deceptions are structural in character; the animal has become a walking (swimming, flying) lie: the praying mantis that stands like
a dry stick, the angler fish whose luminous uvula lures its small victims
right into its jaws, the moth whose markings simulate a bitter-tasting

relative. Higher animals practice behavioral deception. They creep and hide, stalk from downwind, simulate injury to lure the predator from their vulnerable nesting place, double back on their own trails, freeze and then pounce. But a lie, *Webster* tells us, is a falsehood acted or uttered with the intention to deceive, and we would commit the sin of anthropomorphic inference if we attributed intentionality to the mother quail running from her nest dragging one wing. The liar intends to communicate a false belief to his victim, and this intention presupposes a rule-governed system of veridical communication between them. To lie is to break those rules and lying works because the rules are usually obeyed.

To understand why we are such frequent and fluent liars, it is helpful to adopt the perspective of the new science of evolutionary psychology. Just as humans' anatomical and physiological characteristics represent a distillation of those ancestral traits that conferred the most survival value, so too many of our social (and other) behavior tendencies may have resulted from the evolutionary sieve. For our hominid ancestors, the ability to mislead or deceive probably carried an evolutionary advantage just as did fleetness of foot or the possession of an opposable thumb. George Steiner puts it this way:

> Fiction was disguise: from those seeking out the same waterhole, the same sparse quarry, or meager sexual chance. To misinform, to utter less than the truth was to gain a vital edge of space or subsistence. Natural selection would favor the contriver. Folk tales and mythology retain a blurred memory of the evolutionary advantage of mask and misdirection. Loki, Odysseus, are very late, literary concentrates of the widely diffused motif of the liar, of the dissembler as flame and water, who survives.[1]

Ann and David Premack, psychologists who devoted many years to teaching a language-like system of communication to the chimpanzee, made an effort to induce their chimps to lie.[2] The animals know how to tell their trainer where to find a banana hidden outside the cage and they can also tell her that they would like to eat the banana when she finds it. Confronted now with an unreliable trainer, one who they know from past experience is likely to eat the banana himself, will they misdirect his search or pretend that there is no banana? At last report, these honest chimps had not yielded to prevarication in spite of their contaminating contact with humankind. They hide their hands behind them, striving painfully not to communicate at all, but that is as far as they will go. Yet one knows that this laboratory Eden has its serpent, that this innocence cannot endure. From language to lying, the primrose path is short and ineluctable.

If humans learned to lie not long after they acquired language, we may assume that the first attempts at lie detection soon made their appearance. In this respect, ontogeny does recapitulate phylogeny. The child learns first to talk, then to dissemble, then to detect when his or her own leg is being pulled. We are all human lie detectors; we must be to survive in our mendacious society. Those who are least skillful in this respect we call gullible or credulous. Our slang is rich with names for these trusting innocents; we call them cullys, cat's-paws, dupes, greenhorns, gulls, jays, mugs, pigeons, and there are plenty of them—one born every minute. It is interesting that there seems to be no English term for the other end of this dimension, the unusually skillful lie detector. (There is a German word, *Menschenkenner*, that comes close: one who can see into, and perhaps see through, other people. "Skeptic" implies a biased attitude rather than a discerning separation of the false from the true; adjectives like "perceptive" and "astute" are too broad.) If language reflects a society's interests and concerns (as their many words for "snow" reflect the Eskimos' preoccupation with that substance[3]), then this plethora of names for the inept human lie detector—synonyms for "sucker"—may confirm that we are a deceitful and predatory race.

Early societies evolved a variety of exotic methods for getting at the truth, reviewed in detail by several authors, notably J. A. Larson.[4] Torture has always been a favorite technique. Statements uttered on the rack or at the point of death were given more credibility than they deserved, an attitude retained in the common law that attaches special weight to dying declarations. Trials by ordeal generally rested on a religious or magical premise. The innocent would be stronger in combat, the truthful person would be helped by the gods to hold his arm longer in boiling water, or the bleeding from a ritual incision would stop more quickly for the honest man than for the liar. A method worth the attention of modern marriage counselors was employed in medieval Germany to settle allegations of infidelity. The man's left arm was bound behind him and he stood in a tub sunk waist deep in the ground, a short club in his right hand. The woman wore only a chemise having one sleeve longer than the other and in this sleeve was sewn a rock. The procedure required that the wife dance about the tub attempting to brain the husband with the rock while at the same time avoiding the swings of his club. Since it is not recorded how the outcome of this contest was supposed to settle the allegations brought before the court, it may be conjectured that the method was intended to discourage litigation altogether.

In other ordeals, the stress imposed was more psychological than physical. The ancient Hindus required a suspect to chew a mouthful of rice and then attempt to spit it out upon a leaf from the sacred Pipal tree. A man who successfully spat out the rice was considered truthful, but if instead the grains stuck to his tongue and palate, he was adjudicated guilty. During the Inquisition, the Roman Church adapted this technique for testing the veracity of clergy. Instead of rice, barley bread and cheese were placed on the altar in front of the suspect priest and prayers were offered that, should he be guilty of the offense charged against him, God should send His angel Gabriel to stop his throat. Thereupon the priest attempted to consume the bread and cheese. According to Mackay's *Memoirs of Extraordinary Popular Delusions*, "There is no instance upon record of a priest having been choked in this manner." These alimentary techniques have often been identified as precursors of modern lie detector methods, depending as they do on the subject's faith in the procedure, on his fear of being found out if he is lying, and on the physiological reactions that this fear may bring about. In the rice test, for example, fear leads to activity of the sympathetic branch of the autonomic nervous system controlling salivation. The saliva diminishes in volume and becomes viscid in consistency; the rice sticks in the mouth.

The ancients made use of other physiological indicants—sweating, blushing, trembling, the unsteady gaze, the racing pulse. Lombroso, the 19th-century Italian criminologist, and his student, Mosso, adapted the plethysmograph, a device for measuring changes in the volume of a limb, to produce continuous records of pulse and blood pressure changes in their subjects' very similar to those obtained on the "cardio" channel of a modern interrogation polygraph.[5] A. R. Luria, the distinguished Russian psychologist, measured finger tremor and reaction time during the interrogation of criminal suspects.[6] Sir Francis Galton experimented with the technique of word association; when asked to associate to words presented orally, guilty suspects were inclined to show disturbances in the speed or nature of their responses to words related to their crime.[7] But instrumental lie detection—polygraphic interrogation—is a 20th-century phenomenon and as American as apple pie.

William Moulton Marston

A Harvard professor of psychology, Hugo Münsterberg, surveyed all these possibilities in his 1908 book, *On the Witness Stand*, which advocated greater forensic attention to the techniques of experimental psychology. It was William M. Marston, J.D., Ph.D., a student of Münsterberg's at Harvard, whose alleged discovery of a specific lie response marked, in Marston's own modest appraisal, "the end of man's long, futile striving for a means of distinguishing truth-telling from deception."[8] Marston was an avid publicist and either coined the graphic phrase "lie detector" himself or else adopted the expression from one of the reporters to whom he described the wonders of his method. Immediately after the sensational kidnapping of the Lindbergh baby, Marston wrote to Col. Lindbergh, "placing the Lie Detector and my experience with it at his disposal."[9] Lindbergh did not reply, but, undaunted, Marston approached Bruno Hauptmann's defense counsel and the governor of New York with a proposal to test Hauptmann in his death cell as well as "Jafsie," the self-appointed intermediary between Lindbergh and the kidnapper. Rebuffed again, Marston lamented that "the secret knowledge of the crime that Hauptmann had locked in his brain died with him." More positively, "I hope, even yet, to find a living human being whose mind contains information about the Lindbergh kidnapping. If such a person exists, his secret knowledge can be read like print by the Lie Detector."[10]

Marston not only invented modern lie detection but was among the first to appreciate its commercial possibilities. An article in *Look* magazine described his use of the polygraph in marital counseling; a wife's reaction to her husband's kiss is compared with her response to the kiss of an attractive stranger.[11] Marston and his machine also appeared in full-page ads for razor blades verifying that "Gillette shaves you closer!" Finally hitting upon the ideal outlet for his fertile imagination, Marston—under the *nom de plume* "Charles Moulton"—created the comic strip character Wonder Woman with her magic lasso that caused those persons bound in it to tell only the truth.[12]

Because of his extravagant claims and uninhibited conduct, Marston was repudiated by serious students of polygraphic interrogation. In a scathing review of Marston's 1938 book, *The Lie Detector Test*, Professor Fred Inbau concluded, "It can only bring ridicule upon the subject matter and disrespect for its author." Nonetheless, Marston might be called the grandfather ("foxy grandpa"?) of polygraphy because of the early reports

Figure 3.1. Marston starring in an advertisement for razor blades. Source: *Time* magazine, 1938. Courtesy The Gillette Company.

of his work, published prior to 1921, and their impact on Chief August Vollmer of the Berkeley, California, police department and especially on one of Vollmer's police officers, John A. Larson.

Larson and the Berkeley Developments

Larson, who subsequently attended Rush Medical College in Chicago and later became a forensic psychiatrist, began experimenting with the measurement of blood pressure and respiratory changes during questioning. Encouraged by Chief Vollmer, Larson assembled the first continuous-recording interrogation polygraph, and it happened that some of its first practical applications were outstandingly successful. A female college student had been shoplifting in a local store. The shop clerk knew only that the culprit lived in a certain dormitory but could not identify her further. Larson was able to question all 38 residents of that dormitory (in one 16-hour day!) using what later came to be called the relevant/irrelevant, or R/I technique, a mixture of irrelevant questions ("Are you sitting down?" "Is your name Sarah?") and relevant ones ("Have you taken items from Katz's store without paying for them?"). One of the 38 young women responded much more strongly to the relevant questions, as compared to the irrelevant ones, than did any of the other 37, and she subsequently made a full confession.[13]

One can imagine Chief Vollmer's reaction to this episode—a seemingly hopeless case solved as by divination! Not only did the polygraph pick out the culprit from among 38 candidates but the experience so impressed the woman herself that she confessed, sparing further costly efforts, possibly futile, to locate admissible evidence. Larson's reaction was apparently more complex and, indeed, he remains a somewhat enigmatic figure. One gets the impression that he was uncomfortable with words and abstractions, especially so for a psychiatrist. It is difficult to find quotable passages in Larson's writings that reveal clear and succinct statements of his own views. His 1932 book, the first thoroughgoing study of lying and its detection, was a major scholarly effort that brought together in its 400 pages a mass of relevant material, ancient and modern. But oddly, this work is almost entirely quotation, an introductory sentence or two by Larson followed by pages of text from the cited authorities. The difficulty of assessing Larson's own position can be illustrated by considering his Chapter 8, which is concerned with methods of interrogation used by the police of that period. Of the 26 pages in that chapter, at least 20 are devoted to extended quotations from 16 different sources, some piously denying any impropriety in police methods while others catalog examples of confessions extorted by intimidation, threats, starvation, beatings, and confinement in the "sweat box." When he speaks for himself, Larson sometimes appears to defend the police, and yet he ends by saying, "So long as ignorant officials are employed by the police or in the district attorney's office, there will be cases in which a false confession has been obtained by brutal methods."[14] The main point of the chapter seems to be to motivate the use of polygraphic interrogation as an alternative to the brutality that Larson reluctantly admits and obviously deplores.

Another puzzle is why Larson agreed to write a complimentary introduction to Marston's self-aggrandizing potboiler in 1938. In a paper of his own published rather obscurely that same year, Larson insisted that "there is no disturbance graphic or in quantitative physiological terms specific for deception"[15]—that is, that there is no specific lie response, although the opposite claim is Marston's central thesis. Larson devoted a 12-page chapter to Marston's work in 1932, 1 page in Larson's words and 11 of quotation, but he did point out even then that Marston based his approach "upon the assumption, not necessarily always true, that there is an increase of blood pressure during deception." And one wonders how Larson could have condoned Marston's incredible claims about the accuracy of his method, when Larson was at the same time concluding his own 1938 paper with the warning, "Because of the errors of interpretation, and these have been found to be large, a deception test alone should never be used as court evidence."[16]

In view of the boundless enthusiasm and extravagant claims of his associates and successors, Larson's persisting attitude of scientific skepticism seems especially noteworthy. During the first half-century after 1915, when Marston's "lie detector" was becoming entrenched in American folklore, Larson was the only investigator to report an objective study of the accuracy of the diagnosis of deception using polygraph recordings obtained from criminal suspects. In the same 1938 paper quoted above, he tells of examining 62 suspects, whose records were then independently evaluated by nine psychologists. The number of records classified by the different judges as indicating deception ranged from 5 to 33, although 61 of the 62 suspects were in fact truthful. Most recently, Larson has said:

> I originally hoped that instrumental lie detection would become a legitimate part of professional police science. It is little more than a racket. The lie detector, as used in many places, is nothing more than a

psychological third-degree aimed at extorting confessions as the old physical beatings were. At times I'm sorry I ever had any part in its development.[17]

The Chicago Period

It should be emphasized that Larson's experience and at least some of his disillusionment was focused on the relevant/irrelevant test format and that other interrogation techniques, notably the Control Question Test and the Guilty Knowledge Test, were developed later. Until about 1950, the R/I procedure was standard in the field and even now continues to be used by many of the older examiners. It was promulgated especially by two of Larson's associates in the Berkeley Police Department, C. D. Lee and Leonarde Keeler. Lee, formerly a captain of detectives in Berkeley, developed and manufactured a portable field polygraph and wrote a textbook for polygraphic examiners.[18]

Keeler began as a high school student assisting Larson in his early work. He also developed a portable instrument, the Keeler Polygraph, and then in 1930 moved to join the staff of the new Scientific Crime Detection Laboratory of the Northwestern University School of Law in Chicago. That laboratory, which later became a unit of the Chicago Police Department, was an outgrowth of public concern stimulated by the celebrated St. Valentine's Day Massacre in that city. It was in Chicago's bracing atmosphere that polygraphy's seed, planted in Boston, nourished in Berkeley, finally took root and flourished. There Keeler met Fred Inbau, professor of criminal law at Northwestern and later director of the Chicago Crime Laboratory. Inbau was an expert on criminal interrogation and founder of the *Journal of Police Science and Administration*, which, over the years, has published numerous pro-polygraph articles including some by the Raskin group (see below). (Edited by lawyers and police officials, the term "science" in its title should be interpreted as in "domestic science" or "political science.") In Chicago Keeler also met John Reid, another young polygraphy enthusiast who also had his own machine, the Reid Polygraph. It was in Chicago that Keeler and Reid established competing schools of polygraph technique, and substantially all of the polygraphic examiners now practicing can trace their lineage to these two men.

Keeler played an important role both in developing polygraphy and in publicizing the technique. Apparently it was Keeler who invented the card test, used to enhance the subject's faith in the procedure; he also

Figure 3.2. Actor Richard Conte with Leonarde Keeler in a scene from the film *North-side 777*.

developed the important idea of the Peak of Tension Test, which will be considered later. And in the movie *Northside 777*, it was Keeler himself who administered an R/I lie test to Richard Conte in Joliet Prison, proving what reporter Jimmy Stewart had suspected all along, that Conte was innocent of the crime for which he was incarcerated.

Reid's Contribution: The Clinical Lie Test

But a still more important role in the development of polygraphy must be acknowledged for lawyer John E. Reid. His school became the Reid College of Detection of Deception, accredited by the State of Illinois to award a master's degree to the graduates of its six-month course of instruction. His private firm, John E. Reid and Associates, once operated branches in several cities. His textbook with Fred Inbau became the standard in the field.[19] Two of Reid's early students, Richard O. Arther and Cleve Backster, established active schools of their own. Most important of all, Reid developed three new ideas for polygraphic interrogation, the concepts of the "guilt complex" question, the "control" question, and the systematic appraisal of "behavior symptoms."

In a 1947 paper, Reid first described what he then called the "comparative response" question, later elaborated and renamed the "control" question in his textbook.[20] This was a response to the dawning awareness of the inadequacies of the relevant/irrelevant format. After 30 years, polygraphers were coming to realize that questions like "Did you take the money?" or "Did you kill George Fisbee?" must sometimes generate stress or emotional reactions in criminal suspects even when the suspects are innocent and their denials truthful. Reid proposed adding to the list of questions some that were not directly concerned with the crime under investigation but that were calculated to induce an emotional reaction. Specifically, the control questions should be ones to which the subject's answers will probably be untrue. Questions like "Have you ever cheated on your income tax?" or "Have you ever stolen anything?" may be answered "No," and if they are, these answers are unlikely to be literally true. Reid's idea was to use subjects' responses to the control questions as standards against which to measure their reactions to the relevant questions. Backster later provided a psychological rationale for this approach together with other refinements so that Reid's method, in one form or another, has come to be the standard technique used by most modern examiners in dealing with "specific-issue" cases, as in criminal investigation situations.

Reid's proposal to use a "guilt complex" question was, in my view, a much sounder idea than the control question notion, but it is more difficult to implement and never really caught on. In the latest edition of their textbook, even Reid and Inbau treat it as an ancillary technique to be used only in rare and special situations. The idea is, when interrogating about the subject's possible involvement in Crime X, to pretend to be equally interested in whether he might also have been guilty of some fictitious crime, Crime Y. If this fiction can be made convincing, then the subject's response to "Did you commit Crime Y?" provides a good prediction of how that subject ought to respond to the relevant question, about X, if he is being truthful in denying X. Thus, the guilt complex question, properly presented, provides a genuine *control* in the scientific sense of that term, whereas the control questions actually used in every Control Question Test do not. These theoretical issues will be discussed in detail later on in this book and are mentioned here just to illustrate the importance of Reid's role in the evolution of polygraphy.

Reid's advocacy of the use of behavior symptoms as supplementary criteria in the scoring of polygraph examinations has been especially controversial and has been largely repudiated by modern examiners. This proposal grew out of a study, published by Reid and Arther in 1953,[21] tabulating behavior symptoms that, it was alleged, were exhibited during polygraph examinations only by deceptive subjects, together with another list of symptoms said to be characteristic only of the truthful individual. To a psychologist, these tabulations seem at the same time plausible and yet excessively naive. It is plausible that lying subjects might try to postpone their examinations, be late for their appointments, be "very worried and highly nervous" in the examination room, and want to leave promptly when the test is over. But the thought that these and other listed "symptoms of lying" will be systematically toted up and then contribute to the final diagnosis gives one pause. Since the beginning, polygraph examiners had used "global scoring"; the same individual reads the case file, conducts the pretest interview, formulates the questions, administers the test, observes the "behavior symptoms," evaluates the polygraph charts, and then reaches the final diagnosis of deceptive or truthful. Reid's school and especially Arther's emphasized the clinical evaluation of behavior symptoms, insisting that the examiner constitutes the real lie detector, guided by his or her training, judgment, and experience. Yet neither of these two leaders of the global scoring tradition showed any apparent appreciation of the profound psychometric difficulties inherent in their approach, a can of worms I open in a later chapter devoted to the clinical lie detector test.

Backster and the Control Question Test (CQT)

We have already seen that a diminishing proportion of polygraph examiners still employ the relevant/irrelevant test format, whereas some form of Control Question Test is used by the majority. This larger group, in turn, is split into two factions, one of which relies on the Reid–Arther practice of global scoring (also used by adherents of the R/I test). The remaining faction, which includes the most professional of modern-day polygraphers, follows the lead of Reid's other influential student, Cleve Backster.

Backster achieved considerable celebrity some years ago in connection with his experiments with plants. Popular accounts had it that Backster had connected his philodendron to a lie detector, but what he actually did was to clamp a plant leaf between the two electrodes of the electrodermal channel, the part of the polygraph normally used to measure electrical changes in the skin produced by palmar sweating. One can imagine Backster's excitement when the polygraph pen first began to trace out what appeared to be responses from the leaf, changes in electrical resistance not unlike those shown by human subjects. Further experiments persuaded him that these botanical reactions were related to events in the laboratory, even to the experimenter's unspoken thoughts concerning what he planned to do next to the plant! Backster went on to study, in a similar way, the reactions of fertile chicken eggs and, finally, aggregations of living human cells, including spermatozoa. His experiments along these exotic lines generated sufficient interest in scientific circles to lead to an invitation to present his findings at the 1975 meetings of the American Association for the Advancement of Science. Other investigators have been unable to repeat Backster's findings and he never managed to develop an experimental technique robust enough to work consistently in other people's hands.

Backster's contributions to human polygraphy include the "zone of comparison" format, in which only adjacent relevant and control responses are compared so as to compensate for changes in the subject's reactivity over the course of the test. Most important of all, Backster initiated the practice of "numerical scoring" of the polygraph charts. Using rather specific rules and criteria, adjacent control and relevant responses are compared separately for each polygraph channel. Scores are awarded, ranging from $+1$ to $+3$ if the control response is larger and from -1 to -3 if the relevant response is the larger. Summing these scores over all channels and all repetitions of the same question list to get a total score, the test is then classified as deceptive if the total is negative and sufficiently large, as truthful if the total is sufficiently high and positive, or as indeterminate if the total falls in some range about zero. There is no use of behavior symptoms, case facts, or other extraneous considerations; the test outcome is intended to be based entirely on the polygraph findings. Moreover, this systematic numerical scoring procedure makes it less likely that examiners' preconceptions about the subject's honesty will influence their interpretation of the polygraph record. The evidence indicates that competent examiners trained in this method can achieve high levels of agreement when they independently score the same set of charts.

The importance of Backster's innovation lies in the fact that, for the first time, a polygraph examination conducted in this fashion could lay at least tenuous claim to being called a "test" in the accepted sense of that term. A psychological test has four cardinal properties: (1) standardized method of administration (all polygraph methods stretch this requirement since all require questions tailored to the respondent); (2) immediate recording of behavior (the polygraph records immediately, but the examiner's clinical impressions may not jell until the interview is over); (3) objective scoring (the clinical lie "test" is intractably subjective); and (4) external validity criteria (we shall see that there are no clear validity data on the clinical lie test and, what is worse, no really adequate validity studies of the CQT either).

Examinations administered by adherents of the Keeler, Reid, or Arther schools are interrogations, not tests. They are semistandardized methods of clinical observation in which the examiner has the opportunity to collect relevant data. When it is the examiner who is the "lie detector," then it becomes especially difficult to assess the accuracy of the technique in any general way. The Backster technique also depends heavily upon the skill of examiners, especially in their selection of the specific questions to be used. But at least the fruit of the examination, the test results, can be localized in the polygraph charts and we can expect different examiners to score the same charts with reasonably good agreement.

The Boom and Bust of the Polygraph Industry

Beginning in the 1960s, entrepreneurial polygraphers saw the opportunity to transmute their art from a somewhat dubious adjunct to the police forensic laboratory into an independent and profitable industry. Businesses, banks, retail chain stores, fast-food chains, and the like were approached with offers of contractual arrangements by which all entry-level job applicants would be screened by means of polygraph tests designed to determine whether they had told the truth on their employment applications, whether they had stolen from previous employers, whether they used illegal drugs, had outstanding debts, or had undisclosed criminal records. The same companies were informed that instances of internal theft, embezzlement, and misappropriation could most effectively be solved by submitting suspected employees to polygraph tests, and prevented by periodic polygraphic testing of all employees. Large businesses especially lose large sums annually to employee peculation, and these offers of a fast, high-tech solution were attractive. The companies that signed up with local polygraph firms found that a significant proportion of the job applicants they might otherwise have hired were found, by the polygraphers, to be bad risks, and that nearly all of the cases of apparent employee theft that they turned over to the local lie detector mavins were quickly solved—that is to say that in most cases an employee was identified as the culprit and could be summarily discharged.

By the early 1970s, the lie detector industry had grown to the point that at least two million Americans were submitting annually to polygraph testing in order either to get a job or to keep the job they had in the private sector. Polygraph schools, offering courses of instruction lasting six to eight weeks, were springing up around the country. In 1980, when I was collecting material for the first edition of this book, I visited the head of one of the more successful of these firms, Leo Gelb, formerly a lieutenant in the Los Angeles Police Department, who lived up in the Hollywood hills whence he commuted to his office in his new Jaguar roadster. Like most of the polygraphers I've met, Mr. Gelb was a true believer, who was sure that he and his numerous assistants were contributing significantly to the nation's GNP.

Also in the early 1970s, a former rat psychologist (an experimental psychologist who did most of his research with white rats as his experimental subjects) named David Raskin was inspired to spend six weeks of one summer's vacation taking Cleve Backster's six-week course in polygraph testing at his school in San Diego. Raskin at that time was a young professor of psychology at the University of Utah in Salt Lake City. Raskin returned from San Diego a true convert and he began instructing a few graduate students in the techniques of polygraphy. Raskin was not interested in employee screening but he saw a great future for polygraphy in police work, and a lucrative future for scientifically accredited polygraphers as expert witnesses in courts of law.

These two coincident developments had a synergistic effect on polygraphy in the United States. First, employee screening had rapidly become a multimillion-dollar business, and multimillion-dollar businesses, unless they are patently illegal, command considerable respect. And second, polygraphy at last had a respectable, scientifically trained protagonist, someone able to command large fees administering polygraph tests to prominent defendants such as Patty Hearst and John DeLorean, who could conduct and publish apparently respectable scientific studies of the polygraph and who moreover trained and be-doctored at least a half-dozen other young graduate students whose doctoral dissertations were on the topic of polygraphy.

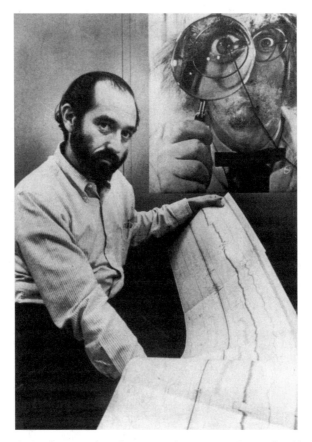

Figure 3.3. Polygrapher David Raskin in *People* magazine, December 12, 1977, insisting that "Patty Hearst told the truth." After viewing photographs of Ms. Hearst participating in a bank robbery, hearing uncontested reports of her firing a machine gun at security guards to effect the escape of her companions, and listening to recordings of her acknowledging her revolutionary views after her eventual capture, a California jury found her to be guilty as charged. Raskin's willingness to rely on his polygraph findings in the face of all this evidence illustrates the strength of his convictions (and also, perhaps, the power of the Hearst family fortune). Source: *People Weekly,* © 1977 John Telford, used with permission.

By contrast, in the mid-1970s, at least one trained scientist, myself, began to try to warn the psychological community about the growth and menace of this malignant industry.[22] On throwing down the gauntlet, I was suddenly inundated with requests to testify in court on behalf of victims of the lie detector and before committees of state legislatures and of the U.S. Congress who were considering some sort of legislative intervention. The antipolygraph crusade came to occupy fully one-third of my professional time. The first edition of this book was published in 1981, the same year that I was elected president of the Society for Psychophysiological Research (SPR), a coincidence in timing that was a great help to this David in confronting the Goliath of the polygraph industry. Before long, other scientists entered the fray. Leonard Saxe, a social psychologist engaged by the U.S. Congress's Office of Technology Assessment to do a study of polygraph accuracy,[23] was shocked by what he found and has become a leading critic of polygraphy,[24] as has a distinguished Canadian psychophysiologist, John Furedy,[25] and an equally distinguished American psychologist, Benjamin Kleinmuntz.[26]

The most important combatant to enter the arena on the side of reason and justice, however, has been W. G. Iacono, Distinguished McKnight Professor of Psychology at the University of Minnesota and, in 1997, also president of SPR.[27] Because Bill Iacono is my former student, it is particularly gratifying to me that he has come to share my interest in the problems of polygraphy and has taken up the challenge of attempting to slay this dragon before it claims yet more victims.

In 1988, happily, an ax fell on the flourishing employee-screening portion of the industry. The federal Polygraph Protection Act prohibited employers in most of the private sector from requiring or even suggesting that prospective employees submit to polygraph testing. To his credit (although he had never profited from the polygraph screening business), Raskin was one of those to testify before Congress in support of this legislation. Hundreds of journeymen polygraphers had to seek other employment and millions of citizens no longer had to face the humiliation of having their character vetted in an hour's time by some graduate of a six-week course in polygraphy. Raskin and his former students, however, continued to flourish. Fees for administering "friendly" polygraph tests to criminal defendants, and then testifying about the findings as an expert witness in those jurisdictions that permit such evidence, now run well into five figures.

Figure 3.4. Professor William Iacono, University of Minnesota.

The Guilty Knowledge Test (GKT)

In 1959, I developed a new method of polygraphic interrogation that was designed, not to detect lying (which I thought then as now to be impossible), but rather to detect the presence of guilty knowledge. The easiest and clearest way to introduce this idea may be by means of an example. In 1986, a Ukrainian immigrant named John Demjanjuk was extradited to Israel for trial as a war criminal. It was alleged by survivors of the Nazi concentration camp at Treblinka that this man had served as the chief guard at that camp and, because of his many atrocities, was known as "Ivan the Terrible." Demjanjuk denied that he had ever set foot in Treblinka, and so, after the changes wrought by so many years, his identification by numerous former prisoners, however positive and impassioned, had to be treated with some skepticism. Imagine administering a CQT to a man in Demjanjuk's situation:

> (Relevant) Were you a guard at the Treblinka camp in 1944?
> No.
> (Control) In the last 20 years, have you ever told a lie to get out of trouble?
> No.
> (Relevant) Are you the man who was known at Treblinka as "Ivan the Terrible"?
> No.
> (Control) Before 1940, did you ever intentionally injure anyone?
> No.

If the suspect was innocent, as he claimed to be, is it likely that he would be more disturbed by those "control" questions than by the (painfully) relevant questions concerning the capital charge on which he'd been extradited?

But if Demjanjuk was indeed the terrible Ivan, his memory would be full of facts, names, places, and incidents that his examiners could learn about from the survivors who had accused him. For example, surely he would recall the name of the chief female guard with whom he worked every day. And certainly he would remember the number of ovens in the Treblinka crematorium, the ones the real Ivan kept busy during his tenure; were there four, or five, or ten? The real Ivan's office was situated across the street from another building; was that the Commandant's office, the guards' quarters, the hospital? Let us collect ten such facts from survivors, making sure not only of their accuracy but also that they would have had to be known by Ivan the Terrible at the time. Let us then construct ten multiple-choice questions, in each of which one of these selected facts will be situated among a group of equally plausible alternatives. Then, in view of the importance of this case, we shall try out our guilty knowledge test on at least one or two people whom we know to be innocent—survivors of other camps, perhaps, who had never visited Treblinka. After connecting the subject to the polygraph, we explain the procedure:

> Now I am going to ask you questions about the Treblinka camp, questions to which you will know the correct answers if you are the man they called "Ivan the Terrible." After each question, I will slowly suggest a series of answers: one of those answers will be the right one and the other answers will be incorrect. I want you to sit quietly and listen to each question and then, when I state the alternative answers, I want you merely to repeat each one after I've said it. Now we shall begin:

1. What was the first or given name of the chief woman guard at
   Treblinka when Ivan was in charge? Was it ... Olga? Greta? Hilda?
   Marie? Marlena? Sophia?
2. How many ovens were there in the crematorium at Treblinka?
   Were there ... Four? Five? Six? Seven? Eight? Nine? Ten?

Our subject is likely to show some involuntary physiological disturbance after each alternative of every question. We examine the polygraph
records to see which alternative in each set produced the strongest disturbance. We never use the correct alternative as the first in a series because
people are likely to respond more strongly to the first just because it is the
first. Suppose that we have six alternatives for each question so that,
ignoring the first, our subject has 1 chance in 5 of giving his strongest
response to the correct answer even if he does not know it is correct. There
is 1 chance in 25 that an innocent subject will "hit" (give his strongest
response to the correct alternative) on both Questions #1 and #2. But there
are only about 3 chances in 10,000 that he will hit on the first five questions,
without guilty knowledge, and only 1 chance in 10 *million* that he would hit
on all ten questions.

We try out our test on innocent subjects to make sure that the correct
answers are not obvious or that some of the incorrect alternatives are not
obviously wrong. Bright college students all can remember doing better
than they should have on badly written multiple-choice tests where the
correct answer could be deduced even by a student who had skipped that
part of the assignment. If our innocent test subjects hit by chance on no
more than two or three of the ten items (and if the ones they do hit on are
different items so that we have no reason to think that one or more of our
items are bad) then we can proceed to test Demjanjuk. If he hits on just six
of the ten items, then we can say that the odds of his possessing guilty
knowledge are about 16 to 1. The odds increase to 256 to 1 if he hits on
seven, and to more than 4,000 to 1 if he gives his strongest response to eight
of the ten questions.

The Guilty Knowledge Test (GKT) has performed very well in a
number of laboratory tests using mock crime situations but it has yet to
be used systematically in real-life criminal investigation. The reason for
this apparent oversight is that professional polygraphers remain persuaded that the CQT in its various forms is highly valid and because the
CQT is vastly easier to use in criminal investigation. The examiner needs to
know only the bare outlines of the charges, does not need to have visited
the crime scene, does not need any real skill in the construction of good

questions. More importantly, there are many situations in which it is simply impossible for the investigator to determine a list of key facts that would be known to the perpetrator but not to an innocent suspect. I have much more to say about the GKT later in this volume.

Voice Analysis

During the 1960s the American military became increasingly interested in the possibilities of covert lie detection and sponsored various R&D efforts in that direction. Lt. Col. Allan Bell, an Army intelligence officer, together with Lt. Col. Charles McQuiston, an Army polygrapher, designed an instrument intended to measure microtremor in the human voice, a low-frequency warble thought to be related to emotional arousal or stress (Figure 3.5). In 1970, Bell and McQuiston resigned from the Army and set up a private company called Dektor Counterintelligence and Security to manufacture their machine, known as the Psychological Stress Evaluator, or PSE. A New Jersey company, Law Enforcement Associates, marketed a competing device called the Mark II Voice Analyzer, which was also developed with Army funding. Both machines cost upwards of $4,000, but one might also consider the Hagoth, a machine that claims to reveal voice stress through a pattern of flickering lights rather than a tracing on a paper chart, and costs only $1,500. More recently, the Hagoth has been replaced by the Truth Phone while the PSE has yielded its place to the Computer Voice Stress Analyzer (CVSA).

Voice analyzers can operate on tape recordings of the subject's speech and therefore can be used covertly, wherever hidden microphones might be employed, and can even be turned to historical research. They have already been used to "show" that Lee Harvey Oswald was truthful when he denied killing President Kennedy and that Bruno Hauptmann was innocent of kidnapping the Lindbergh baby (the researcher here employed a smuggled recording of Hauptmann's courtroom testimony; how envious Marston would have been!) and to assess the veracity of President Nixon, John Dean, and other televised notables.[28] It should be emphasized that none of these voice analyzers are claimed to be able to detect deception directly. Like the polygraph, they are said to measure variations in emotional stress; whether a particular statement is deceptive is inferred from a comparison of the "stress" associated with that and other utterances by the same respondent. It must be admitted that some voice analysts, like some

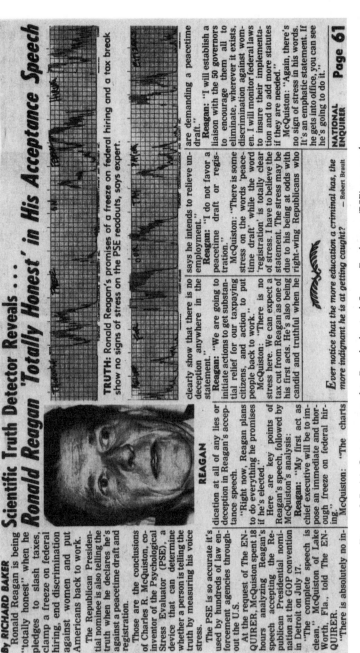

Figure 3.5. McQuiston puts his Psychological Stress Evaluator (PSE) to work.

polygraphers, operate as if lying is the only reasonable explanation for an emotional reaction to a question, thus committing themselves to the simplistic assumptions of the relevant/irrelevant test format. The polygraph does measure variations in nonspecific emotional arousal (except when the subject knows how to produce misleading reactions by means of self-stimulation). The problem with various methods of polygraphic interrogation is whether deception can plausibly be inferred when the subject is more aroused by one question than by another. Voice analyzers are confronted with this problem, as are all other proposed methods for recording stress—those using skin temperature, changes in pupil diameter, even body odors. If a given type of voice analyzer *can* measure emotional arousal, then it could be used for lie detection no less accurately but no more so than the traditional polygraph. As we shall see, the voice analyzers have yet to cross this first hurdle; there is no convincing evidence that any of them even measure stress.

Honesty Tests

During the 1970s and 1980s, the principal application of the polygraph or the voice analyzer was in the screening of job applicants to determine whether they were in the habit of stealing from previous employers or whether they were truthful in answering the questions on their application forms, in short, to determine whether the applicant was likely to be a reliable, trustworthy employee. When most such activity in the private sector was banned in 1988 by the federal Employee Polygraph Protection Act, many employers turned to an alternative approach. This was to administer a paper-and-pencil "honesty test" like the Reid Report or the Stanton Survey or the Personnel Security Inventory, multiple-choice questionnaires that are easy and inexpensive to administer and that are claimed to be able to select honest employees. The Reid Report, the first honesty test on the market and typical of the group, was developed by the John Reid organization of Chicago for sale to those of its business clients who were unable or unwilling to use the more costly polygraph screening. The close connection between the honesty tests and the methods of instrumental interrogation already described is highlighted by the fact that the evidence offered for the validity of these questionnaires consists mainly of correlations with the polygraph. For example, when the Stanton Survey was administered to a group of applicants who were also given a poly-

graph test, nearly two-thirds of the applicants "failed" both tests. I shall
discuss the implications of such remarkable findings in Chapter 16.

The Tools of Diogenes

We can see that the modern Diogenes has many tools at his disposal,
and it is my purpose in this book to examine each of them critically. There
are honesty tests and there are methods of instrumental interrogation. The
interrogation methods may employ a voice analyzer device or a conventional polygraph. One experimental method of polygraphic interrogation
attempts to determine whether respondents possess "guilty knowledge,"
whether they recognize facts, scenes, or objects that they could identify
only if they were guilty of the crime in question and had been present at
the scene. Virtually all instrumental interrogation now used in the field,
however, is intended to detect lying, to determine whether the respondent's answers to certain relevant questions are deceptive or truthful.
Whatever instrument is used, the lie detector examiner may employ one of
several question formats. Each of these question formats requires a different set of assumptions if deception (or truthfulness) is to be inferred from
the polygraphic responses. Whatever format examiners use—and whichever instrument they employ for recording the subject's reaction—they
may arrive at their final judgment either through clinical evaluation or
through objective numerical scoring of the polygraph (or voice analyzer)
records alone. Examiners who use clinical evaluation allow themselves to
be influenced not only by the instrumental findings but also by the respondent's demeanor and behavior symptoms, the case facts, and other sources
of intuitive insight.

After 50 years of virtual neglect, there has recently been a wholesome
increase of critical discussion of this truth technology and its potential
impact on our society. But much of this discussion, both in law journals
and in the popular press, has treated these varied techniques in an undis-
criminating way, as if they were all equally valid or equally absurd, equally
useful or equally pernicious. Among polygraphers, proponents of one
method scrupulously avoid criticizing other methods, at least in print,
perhaps in fear of providing ammunition to critics from without. Although
their business is flourishing, the truth merchants are sensitive to outside
criticism and quickly close ranks in bristling solidarity. Some see polygraphy as in a cold war and adopt the tactics of that period. Richard Arther,
in his text *The Scientific Investigator*, exemplifies this approach:

> Obviously, from our [foreign] enemies' viewpoint the only way to stop
> their exposure is to stop the polygraph. That is the reason for most of
> the anti-polygraph articles in the press and a great deal of the legisla-
> tion being introduced in some of our states by Communist dupes and
> fellow travelers.[29]

Another reason why proponents of the truth technology tend to be
less discriminating than they should be is, I think, ideological. Many
polygraphers and many of their clients are concerned about what they see
as a breakdown of law and order and are tolerant of almost any development intended to help stem this seeming tide of anarchy. Listen, for
example, to polygraphers Ferguson and Miller:

> Permissiveness, lack of conformity to the accepted rules and regula-
> tions long governing moral conduct in this country, has already de-
> prived the order and right of the sincere majority for the sake of
> pacifying a few depraved, warped minds. This is the epitome of mis-
> guided humanitarianism at best. At worst, it approximates social sui-
> cide, twisting liberty into license from law. Due to our legislative and
> court-affected complacency, we have seen our cities convulsed with
> mindless destruction in the name of *civil rights*, universities bomb
> blasted or shut down by self-styled revolutionaries in the name of
> *academic rights*, sedition and draft evasion by cowards and overedu-
> cated milque-toasts in the name of *morality*, militants and radicals
> openly advocating guerrilla warfare and anarchy in order to disrupt
> the democratic process and overthrow the government in the name of
> *freedom*, a youth culture anesthetized and sustained by drugs in the
> name of *self-expression*, hippies and flower children aimlessly wander-
> ing and littering the streets in the name of *peace*, an increase of gonor-
> rhea and syphilis in the name of *free love*, and finally a shambles of
> primitive lawlessness in the name of the *right to equal shares for every-
> body*.
>   This kind of complacency and toleration, with respect to crime
> and moral conduct, is the deadliest of diseases. If a deadly disease is
> not checked, death results—in this case, the death of a nation. No
> matter how hard it is to accept, a democracy such as ours, like less
> libertarian systems, cannot grant all freedoms to individuals and leave
> none for itself to use in the interest of its own preservation.[30]

Not all polygraphers share this menacingly authoritarian outlook. But
the great majority of them come from a background in police work or
military intelligence and they are not always ardent supporters of those
various rights and freedoms that Ferguson and Miller deprecate.

Opponents of the truth technology tend to emphasize issues like the
right to privacy and the dignity of man and seem to take the position that
all these techniques should be proscribed on the basis of the same set of

libertarian principles. Thus, Robert Ellis Smith, editor of *Privacy Journal*, asserts,

> Even if polygraphs were regarded as totally reliable, I would still oppose their use as lie detectors, just as I oppose the use of wiretaps. Wiretaps, after all, are totally reliable, but they still violate individual privacy.[31]

Note that this is a false analogy. When he refers to the "reliability" of the polygraph, Mr. Smith has in mind the accuracy of the lie detector test. Because it is not a test, a wiretap can be neither accurate nor inaccurate in that sense. Or consider the splendid fulminations of law professor E. A. Jones:

> If the lie-detecting polygraph were indeed to be what it is *not*— monument to technological infallibility—if it were a chrome-plated, flickery-lighted, super-efficient computerized conduit of discovery, linked to the sweaty wrist and breath-gulping, heaving chest of an evasive, guilt-worried, fault-smothering, self-excusing human being, which is surely descriptive of what each of us has become on some occasions in the course of our lives, I would still come down on the side of exclusion. Each of us is too imperfect and fragile a creature to sustain such rigorous thrusts of suspicion and rejection into our being and yet maintain that sense of personal worth and higher purpose—and recurrent resolve to do better—which is indispensable to dignity and accomplishment. l think it is far preferable that a fellow human, concededly imperfect in the capacity to perceive calculated falsehoods, be the assessor of credibility than to achieve a mechanical perfection akin to Orwell's *1984* and Huxley's *Brave New World.*[32]

Although I respect both these writers and share their values and concerns, I believe that their approach is mistaken both substantively and tactically. This book is anti-Luddite in its conviction that these tools of Diogenes will be used if they work and that they cannot be effectively opposed merely because they are new or employ the trappings of technology. If there is or could be a method to infallibly detect lying, then I shall argue that such a resource would be irresistible, that it would come to be used widely and would produce a social revolution to which these principles of privacy and dignity would have to accommodate themselves. A method of lie detection that is invalid, based on deceit, or tainted by examiner bias can more confidently be opposed for those reasons than on the basis of abstractions with respect to which public opinion is widely divided, whatever the Constitution may be interpreted to say. Many Americans, for example, do not really object to warrantless search by the police (until it happens to them) because "if you have nothing to hide, you have nothing to fear." Such citizens are unlikely to support a ban on lie detectors because they "intrude into man's interior domain"—the basis on which Pope Pius XII condemned the lie detector in 1958[33]—but they probably would support a ban if the lie detector can be shown to be biased or inaccurate.

Between these two extremes of chance accuracy and near infallibility there remains a wide range to be discriminated. If an instrumental method for affixing guilt could be shown to be, not infallible, but able on the average to make only half as many errors as our criminal courts make now, who would want to reject that method out of hand? In assessing credibility or predicting honesty, certain methods such as physical abuse or intimidation are intrinsically intolerable, but barring these, is not the accuracy of the method more important than its nature? Employers will attempt to predict the trustworthiness of a job applicant. Is it better that they should do this intuitively from interviews and background checks even if a questionnaire or a brief, standardized polygraph interrogation could be shown to be objective, unbiased, and more accurate? I do not mean to argue that questions of privacy, human dignity, or constitutional principle are less important than questions of the accuracy and reliability of this new technology. On the contrary, my thesis is that these higher principles are better served if their consideration is deferred until one is quite clear about what these various techniques are like, how accurate they are in identifying the culpable, what price is paid in misidentifying the innocent, and thus how strong a case can reasonably be made in favor of their use.

It will therefore be apparent that this book is not written by a lawyer. I once examined a 15-year-old boy who had been adjudicated guilty of murder by a juvenile court and referred for presentence evaluation. Convinced that this lad was psychologically incapable either of the killing itself or of behaving as he did afterwards if he had been guilty, I studied the state's case against him, even going so far as to hire a private investigator to look into the matter, and then explained in my report to the court why I believed that the facts denied his guilt. After being severely admonished by the judge for "meddling in evidentiary matters," I came to realize that the legal mind and the scientific mind differ in subtle and unexpected ways, their diverging concepts of the nature of (and professional claims upon) "evidence" being only one example. Judges who scrupulously administer the established rules of evidence can be astonishingly credulous in respect to certain psychological inferences. For example, Judge Jerome Frank reveals

absurd rules-of-thumb some trial judges use, such as these: A witness
is lying if, when testifying, he throws his head back; or if he raises his
right heel from the floor; or if he shifts his gaze rapidly; or if he bites his
lip. Every psychologist knows how meaningless as signs of prevarication any such behavior may be.... Not very long ago, a federal judge,
toward the end of his long career on the bench, publicly revealed for
the first time that he had always counted as a liar any witness who
rubbed his palms while testifying. That judge must have decided
hundreds of cases in which he arrived at his [findings of fact] by
applying that asinine test for detecting falsehoods.[34]

An article in the *New York University Law Review* begins by explaining
that the authors do not intend to consider the "scientific reliability of
polygraph evidence" but will rather assume "that the lie detector test
results are scientifically reliable enough for some evidentiary purposes."[35]
They then proceed with a 30-page discussion of precedents, constitutional
issues, and, surprisingly, a section on "Weighing the Probative Value of
Polygraph Evidence," although how one can usefully discuss the "probative value" of a test without considering the evidence for the validity of
the test is hard to understand for one not learned in the law.

A psychologist, therefore, especially admires those legal scholars who
seem to think like scientists. Attorney Lee M. Burkey[36] and law professor
Jerome H. Skolnick[37] are highly recommended for this reason. Another
law professor, Edgar A. Jones,[38] has provided an analysis of the psychology of both the polygraph examiner and the examinee that is brilliantly
perceptive. Enough of his admirable prose will be quoted herein to tempt
the reader to read Jones's chapter, intended for contract arbitrators and
other triers-of-fact, but accessible to any interested person. I was especially
impressed by a Canadian Supreme Court justice, the Honorable Donald R.
Morand, who was appointed in 1975 to preside over a Royal Commission
to investigate the practices of the Toronto metropolitan police. Polygraph
evidence had been offered concerning the veracity of various witnesses in
this matter, and noting that American enthusiasm for polygraphy was
tending to drift north across the border, Justice Morand determined to
seize this opportunity for a thorough study of the matter. Many days of
testimony, composing 11 volumes of the commission record, were obtained
from leading polygraphers and a number of interested scientists. A full
chapter of the commission's 1976 report is devoted to an appraisal of this
record.[39] Inevitably, Justice Morand looks first at legal precedents, but it
becomes quickly apparent that he is primarily concerned with the concrete
nature of these techniques, the plausibility of the assumptions on which

they are based, and the independent evidence available concerning their reliability. Some of this judge's trenchant observations will be quoted in the following pages.

Science and the Lie Detector

None of the major figures in the development of instrumental interrogation methods have had credentials as scientists. The Backster form of polygraph examination is a psychological test. Physiological reactions are used as a basis on which to draw inferences about the psychological state of the respondent: Was he more aroused by this question than by that one? Was he lying? The Reid or Keeler forms of examination are not strictly tests but their objective, similarly, is that of making a psychological diagnosis. And yet Reid and Inbau are lawyers, not psychologists; Keeler, Backster, and Arther had no professional psychological or scientific training; Lee and even Larson, during the main period of his work in polygraphy, were policemen. At the present time, perhaps ten of the thousands of practicing examiners might meet requirements for licensure as consulting psychologists. In view of the extravagant and discredited claims of W. H. Marston, with his Ph.D. in psychology from Harvard, it might be held that this conspicuous lack of participation by professional psychologists in the development of polygraphy has been all to the good.

The creator of Wonder Woman notwithstanding, the fact that this major area of applied psychology has such tenuous roots in psychological science is cause for concern. There is a great body of knowledge and theory concerning psychological testing in that branch of psychology known as psychometrics or psychological assessment, a database with which only a handful of polygraphers are able to make contact. The use of physiological responses as indicators of psychological states or events also is a subdiscipline of academic psychology, the field known as *psychophysiology*. Again, only a handful of polygraphers have had significant basic science training in this area. Psychologists specializing in psychometrics or in clinical psychology know something about the perils and pitfalls of psychological assessment, the factors that diminish test accuracy, and they understand how to measure the reliability and the validity of a test in an objective and credible fashion. Academic psychologists know (or should know) how to conduct, report, and interpret scientific research.

Finally, and this is especially important, academic psychologists maintain the archival literature of their discipline, scientific journals in which appear only those research reports that have survived a screening process of peer review. If I wish to publish an article about lie detection in the *Journal of Applied Psychology* or in *Psychophysiology*, my contribution must first be read by the editor and by at least two editorial consultants, scientists knowledgeable about the substance and methods involved in my work, and my article will not be published unless or until these individuals are satisfied that it makes both a valid and a useful contribution to knowledge. This screening process is by no means infallible. I have had brilliant(!) papers rejected by pig-headed reviewers, and other researchers have managed to slip flawed or trivial manuscripts past the half-closed eyes of sleepy or incompetent editors. But the scientific peer-review process is nonetheless enormously important, as can be inferred from the fact that 60% to 85% of the papers submitted to these journals are rejected altogether and a high proportion of the remainder appear in print only after more or less extensive revision to remove ambiguities, overstatements, and downright errors detected by the editors.

People are inclined to think that books are more substantial and authoritative than mere journal articles. A book is undeniably harder to write (as I have discovered toiling at this one), but even scientific books are seldom subjected to really serious peer review before publication. The handful of books on polygraphy (this book has in fact been screened by a number of experts) represent the view of the author, screened only by himself, and the reader must be his or her own judge of the logic of the arguments and the quality of the evidence presented. A substantial proportion of the literature of polygraphy over the past 40 years has appeared in law journals, especially the *Journal of Criminal Law, Criminology and Police Science* (of which Reid's collaborator, Professor Inbau, was for many years managing director and then editor), in police journals such as *Police*, *Journal of Police Science and Administration*, *Military Police Journal*, or *Law and Order*, or in trade journals like *Polygraph*, *Journal of Polygraph Studies*, *Security World*, or *Banker's Monthly*. These all may be perfectly respectable publications of their type, but they are not scientific journals and none of them requires their contributors to pass the scrutiny of sharp-eyed scientific editorial review. For someone attempting, as I am here, a fair and reasonably comprehensive critical survey of this literature, the vast bulk of which has not had such preliminary screening, this situation is a kind of nightmare. I would expect that anyone with a scientific background who attempted to read critically all of the 1,700 references cited in Ansley and

Horvath's bibliography of polygraphy, *Truth and Science*,[40] might risk serious psychiatric consequences. I have read widely but selectively, emphasizing the scientific literature, the major books in the area, many of the law journal articles, plus *Polygraph*, the official journal of the American Polygraph Association.

The scientific literature relating to polygraphy is small and only a handful of scientifically trained investigators are currently active, notably W. G. Iacono, at the University of Minnesota, G. Ben-Shakar, S. Kugelmass, and I. Lieblich at the Hebrew University of Jerusalem, and A. Suzuki and associates in Japan. Except for F. Horvath, at Michigan State University and formerly with the Reid firm, most academic proponents of polygraphy were trained by David Raskin, at the University of Utah. They include G. Barland, at the Army Polygraph Institute in Alabama, C. R. Honts, at Boise State University, S. Horowitz, at Central Connecticut State University, and J. Kircher, at the University of Utah. Horvath, the Raskin contingent, and the Japanese group are practicing polygraphic examiners, a fact that helps to explain their interest in these problems. Fewer than 0.03% of the accredited psychologists in the United States have contributed to the polygraphy literature. Yet polygraphic interrogation is unquestionably one of the most important areas of applied psychology in the United States, measured in terms of its wide social impact. And polygraphy is an interesting and challenging problem scientifically as well as in its public policy aspects. It seems to me that this neglect by psychologists and psychophysiologists is unfortunate, baffling, and somewhat scandalous.

The lie detector is currently used in criminal investigation and security applications in Canada, Israel, Japan, South Korea, Mexico, Pakistan, the Philippines, Taiwan, and Thailand. Many foreign examiners have been trained at the Army Polygraph School at Fort McClellan, Alabama. Employee screening applications were extensive in the United States until the federal Polygraph Protection Act of 1988 forbade their use in the private sector, but ironically, federal agencies as well as local police departments continue to use polygraph screening. Western European police agencies remain skeptical about the value of the polygraph. Scotland Yard, for example, has reviewed the matter and concluded that the polygraph does not fit in with its methods and policies. Like Coca-Cola and the snowmobile, the lie detector has become a distinctive feature of contemporary American culture. We shall decide later whether this fact is a reason for chauvinism or chagrin.

Chapter 4

THE TRUTH VERIFIER

To unmask falsehood and bring truth to light.

—Shakespeare

No normal human being wants to hear the truth.

—H. L. Mencken

It seems natural to believe that a machine can detect lying, a machine not so simple as a smoke detector, perhaps, but more like the cardiac monitor connected to intensive care patients in hospitals that turns on an alarm when heart action strays beyond normal limits. The familiar term "lie detector" conjures such an image, a machine that rings a bell each time the subject tells a lie. A moment's reflection makes clear that such a device could exist only if everyone produced some distinctive physiological response when attempting to deceive, a response never produced, for example, when telling the truth but feeling frightened or afraid that he or she may not be believed. As Dr. Marston correctly pointed out, "It is necessary to test for some emotion which will not be present unless a person is lying ... some *one* bit of behavior which would *always* mean a person was lying. Early in the twentieth century this long sought symptom of deception was discovered."[1] As we shall see, Marston's claims of discovery were premature. But if there were such a specific lie response, then modern psychophysiological techniques would probably allow us to detect it and the dream of a genuine "lie detector" would be a reality. It will be instructive

to consider how society as we know it might be changed by the existence
of such an instrument.

The Dawning of the Age of Truth

In 1961, a conference was convened in Washington, D.C., by the Institute for Defense Analyses (IDA) to discuss potential contributions of the
polygraph to national security. The Institute for Defense Analyses was a
kind of think tank operated by the U.S. Department of Defense. Attending
the conference were most of the American scientists then living who had
done research on polygraphic interrogation, a total of eight people altogether. One of the topics for discussion was whether use of the lie detector
might not be incorporated into the nuclear test ban agreement then being
negotiated with the Soviets. In those days the skies were not yet filled with
Russian and American spy satellites, and the physical scientists could not
promise to detect more than 20% of any nuclear weapons testing done in
violation of the treaty. Perhaps each side could be permitted to administer
polygraph tests at regular intervals to key officials of the other government? But this raised the possibility that the Soviets (or we) would keep
one set of figurehead officials ignorant of actual test ban violations or that
knowledgeable persons might be made immune to lie detection by hyp-
nosis, drugs, or special training. We (and the KGB also) later discovered
effective countermeasures to defeat the lie detector, but in the meanwhile,
the satellites began to fly and the urgency of this need dissipated.

Another proposal mooted at this IDA conference by a distinguished
neurophysiologist[2] was vastly more ambitious. First, the standard field
polygraph, a relatively primitive device reflecting 1930s technology, would
be modernized, transistorized, and otherwise brought up to date. Then it
would be rechristened; the lie detector would become the Truth Verifier.
Finally, redesigned and renamed, the instrument would be employed at
the highest levels of international diplomacy. We were asked to imagine
then-president Kennedy or Chairman Khrushchev addressing the General
Assembly of the United Nations. The president or chairman, on the po-
dium confronting all the world's representatives and TV cameras, would
be connected to the Truth Verifier; the Verifier, in turn, would be connected
to a huge meter positioned on the wall over the speaker's head. As long
as each national leader confined himself to what he believed to be the
truth, the meter-pointer would bounce along comfortably near zero. But

should he assert anything about his nation's actions or intentions that he knew to be misleading or deceptive, the pointer would then swing revealingly over into the lie zone.

One's first reaction to such a scenario is mocking incredulity; this is science fiction. But if there is a specific lie response—like the lengthening of Pinocchio's nose—then we can surely design an instrument to measure it. And, if we had such a Truth Verifier, almost infallible (nothing in the real world is wholly infallible), why not use it as a specific antidote for the international suspicion and distrust that constitutes the most important single threat to world peace? Deception is so pervasive in human intercourse that it is difficult to imagine a world in which dissembling had become impossible. How far would Hitler have been able to pursue his plans, prior to the invasion of Poland, if all the world had known *for certain* what his true intentions were from the time he became chancellor? Surely France, Britain, and the Soviet Union would have mobilized in pace with Germany to meet the threat. The surprise attack on Pearl Harbor with its enormous military advantages for the Japanese could not have happened if diplomacy had been conducted with the help of a Truth Verifier. Moreover, it is unthinkable that American public opinion, however reluctant and inner directed, would have failed to support military preparedness if shown unequivocal proof of Japanese intentions as early as 1938.

Diplomatic mendacity is not always an essential precursor to hostilities, of course. A stronger nation might openly decide to attack a weaker neighbor that had neither friends nor resources adequate to mount a serious resistance. In earlier times well-matched rivals frequently engaged each other for emotional reasons rather than for coldly calculated material advantage. Street fights sometimes occur precisely *because* the combatants know the truth about each other and not as any consequence of lying. But the nuclear threat to world peace is thought to be of a different order. It was inconceivable that the Russians and the Americans would push the buttons of mutual annihilation because of injured pride or dreams of glory. It was generally agreed that neither side would strike if it was convinced that the opponent would retain the capacity for devastating retaliation. But both sides feared that the opponent might develop technology of offense or defense that would unbalance the deterrence—and therefore both sides worked and spent feverishly to accomplish just such an imbalance! But suppose both sides knew *for certain* not only the current intentions, but the other's military capabilities as well. Suppose further that all weapons research was, in effect, conducted publicly with its fruits immediately available to both sides. What would be the risk of nuclear holocaust then?

To accomplish such a happy reduction of mutual fears and suspicion, the Truth Verifier would have to be used carefully, and both sides would be required to permit free access to each other's military planners and technicians. Unaccustomed to operating in a context of openness and truth, both sides undoubtedly would at first shy at the thought of exchanging all their secrets, no matter how securely verified as to completeness and accuracy. But the benefits of such an exchange, on a continuing basis, would surely overwhelm these initial doubts. Not only would the danger of war be reduced, but also the great weight of military expenditure. Both parties to such an agreement would doubtless continue to study possible methods of defeating the Truth Verifier, for fear that if such a method did exist, the other side might discover it first, but each side's progress in this respect would be fully known to the other. In contrast, expensive and dangerous weapons research might be expected to receive less and less priority. Why spend billions to perfect a laser death ray when your prospective enemy is all the while looking over your shoulder—and not spending anything? The stalemate of deterrence could presumably be maintained with a fraction of the nuclear weapons that were currently deployed. Substantial disarmament might finally be achieved once both sides could be given verified assurance of the other side's compliance.

Having accomplished disarmament and world peace, what other changes might the Truth Verifier work upon our cynical society? One thinks first of the law courts, most of whose time is presently devoted to the toilsome and often frustrated effort to determine the truth about what really happened and who did what to whom. While conflicting testimony does not always indicate that somebody is lying, our courts would be transformed beyond recognition if perjury were no longer possible. Juries would certainly become redundant in criminal courts and in most civil cases as well. To find out if the defendant is guilty, if he in fact did what the prosecution alleges he did, we would have only to ask him, with the Truth Verifier calibrating his reply. Such a radical change would no doubt be opposed at first. To require a defendant to plead to the charges while being monitored by the Truth Verifier would be said to be in violation of Fifth Amendment guarantees against self-incrimination. But no one is happy with the present system in which the guilty too often get off, especially if they can afford the best lawyers, while the innocent too often go to jail, especially if they are poor.

If the Verifier could arrive at the truth in just hours, when the present system, often after weeks of costly argument, goes frequently astray, who can doubt that the Fifth Amendment would eventually be interpreted so as to permit the use of the fairer and more efficient method? The respected legal scholar John Henry Wigmore, whose *Treatise on Evidence* is known to every lawyer, once suggested having a lie detector in the courtroom in a position where the jury could see the readings as they were made—recorded electronically as the witness testified.[3] The Defense Department did not invent the Truth Verifier after all!

Harold J. Rothwax has practiced criminal law in the borough of Manhattan for more than 40 years, as a defense attorney and then, for the past 25 years, as a trial judge in the New York State Supreme Court. In an alarming recent book,[4] Judge Rothwax tells us that the grotesquely overburdened courts of his jurisdiction are able to provide the constitutional right of a jury trial to only about 1% of the 112,000 persons arraigned each year for minor crimes, and to less than 10% of the 13,000 defendants indicted for felonies. The result is that more than 90% of the people arraigned in Manhattan for crimes ranging from purse snatching to multiple homicide must have their fates decided by plea bargaining. A very few of these defendants are actually innocent, yet, on the advice of counsel, they plead guilty to some reduced charge rather than to face what Judge Rothwax himself calls the "crap shoot" of a jury trial that might lead to a much harsher penalty. Most of these plea bargainers, however, are guilty, and the result of this system is for them to escape the punishment appropriate to their true offense and, in many cases, to put dangerous felons back on the street, where they will rob and kill again, much sooner than if the system had the resources required to convict them properly before a jury.

Jury trials in Manhattan engage the full-time efforts of a judge, two or more attorneys, 12 jurors plus alternates, and a full suite of clerks, bailiffs, and other personnel for an average of some two weeks each. Judge Rothwax does not provide an estimate of the typical per diem cost of such a trial, but I am sure he would agree that $3,000 per day or about $60,000 per trial is surely a conservative guess. In two weeks' time, however, any competent polygrapher could administer careful Truth Verifier tests to at least ten defendants. If we allow for his and a secretary's salary plus office costs, that works out to about $1,000 per defendant, or a savings of more than $1 million per polygrapher each year! Put somewhat differently, just six full-time polygraphers could easily adjudicate all 1,350 cases now being tried annually in the 20 Manhattan criminal courtrooms and the 40 New York State Supreme Court trial parts. More importantly, it would require a staff of only 52 polygraphers to adjudicate all 13,000 of the alleged felons who are indicted annually so that none of the guilty would escape their just deserts.

The Truth Verifier would deter crime not only by ensuring the conviction of apprehended criminals, but also by increasing the likelihood of their apprehension in the first instance. Who knows how many criminals each year are questioned by the police and then released for lack of probable cause? It is estimated that 95% of felonious crimes in the United States remain unsolved[5]; that shocking figure, itself in part responsible for our high crime rate, would certainly be lowered if every suspect were to be questioned under the scrutiny of a Truth Verifier. In any situation where one now may be required to swear or sign an affirmation of veracity, the use of the Verifier would be commonplace. Income tax evasion would dwindle to insignificant proportions; so would welfare fraud, forgery, contractual misprision, voting fraud, misrepresentation of identity, citizenship, or other credentials, and deception in employment applications. Ultimately, to the consternation of civil libertarians, the notion of periodic Verifier screening of all citizens would be bound to be suggested. Just the one question, "Since your last Verifier screening, have you committed one or more felonious crimes?"—those who ring the Verifier's bell to be interrogated further. An indignity, perhaps, and certainly a small nuisance, but what would it be worth to convince the would-be criminal that the probability of his being caught and punished for his crime was, say, 99% rather than the present sporting figure closer to 5%?

The political arena would be hardly recognizable in the Age of Truth. A Truth Verifier would certainly be kept in the White House auditorium where the presidential press conferences are held, this instrument to be calibrated and maintained by the National Bureau of Standards. It seems likely that all the leading news agencies and wire services, the major newspapers, and the broadcasting networks would equip their reporters with mini-Verifiers for on-the-spot interviews. Congressional oversight committees could for the first time genuinely monitor the actions of the CIA and the other investigative agencies. Senior bureaucrats could be held strictly accountable for their departments since, because of the Verifier, they could no longer reasonably plead ignorance of the misfeasance, malfeasance, or nonfeasance of their subordinates. The Truth Verifier would be a cross that the seekers of elective office would have to learn to bear. A considerable adjustment would be required, not only of the politicians, but also on the part of the electorate, before we could all live comfortably under a system where everyone knew what the candidates really thought about the issues—and also which issues the candidates really knew nothing about.

It is but a short step from political campaigns to the world of marketing in general. Some of the earliest models of the Verifier would quickly appear in television commercials, the president of some company whose product really is better than the competition's staring woodenly into the camera and declaring that "Gumgrip holds your dentures tighter!"—with the Verifier by his side to prove it. Do not imagine that we have descended into trivialities when we think about the advertising applications of the Verifier. Think of the impact on the producers of inferior products. In the Age of Truth, they could not advertise at all. As everybody knows, except the spokesmen for the Advertising Council, billions are spent each year to persuade us that inferior or overpriced products are not only desirable, but cheap at the price. Under the influence of the Verifier, those billions would have to be invested instead in research and development aimed either at producing a better product or at least a cheaper one, so that the company could get back on the air again.

The impact of the Truth Verifier on everyday human relations is harder to forecast. If we are to accept Professor Sissela Bok's analysis,[6] to lie to someone, even with the best intentions, is to commit a form of bloodless violence upon them, ethically acceptable only in the same rare and special circumstances where physical violence might be justified (for example, concealing someone's whereabouts from an evil pursuer) or where one can be certain that the harm done by the truth will substantially outweigh the injury of deception. (I do not give an example of this situation because I cannot think of one; the classic example, in which the physician hides the grim truth from his patient, is the one Bok's book was written to refute.) My own personal view, for what it is worth, is that face-to-face lying, as distinguished from institutional deception (where either the liar or the lied-to or both are institutions rather than individuals) is a lot less frequent than we normally suppose, that most interpersonal falsehoods are committed in the heat of passion or in the murk of self-deception and do not involve a clear intention to deceive. In any case, since the Truth Verifier *is* science fictional after all, we may as well decree that it will never be usable covertly, without attachments to the body of the person monitored, so that it will not become a standard household gadget governing our everyday relationships with family and friends.[7]

Is There a Truth Verifier?

We have already noted that for there to be an instrument that literally detects lying, there must be a specific lie response, a distinctive, involuntary, bodily reaction that everyone produces when lying but never when telling the truth. There will never be a Truth Verifier (lie detector) until a specific lie response can be identified. Subjectively considered, it seems plausible that lying might involve a characteristic physiological accompaniment; we all know a "guilty feeling" when we experience it and most of us usually feel guilty when we lie. Guilty feelings must involve different internal states than do feelings of fear or anger or delight, since otherwise we could not distinguish one from another. Could the specific lie response consist of that involuntary psychophysiological reaction that accompanies the feeling of guilt?

The subjective experience of the various emotions is so vivid and distinctive that it is natural to assume that each emotion is associated with a unique pattern of autonomic response, to believe with Franz Alexander that "every emotional state has its own physiological syndromes."[8] But the evidence for this assumption is surprisingly scanty. In a classic study done in 1953, Albert Ax employed a complex laboratory polygraph to monitor some dozen different physiological variables in his subjects while they were hoodwinked by elaborate stagecraft designed to make them either angry (Group A) or frightened (Group F).[9] On the average, fear and anger produced different patterns of physiological reactions. J. Schachter repeated Ax's study in 1957, adding a third group, who experienced the dull, aching pain of a "cold presser" test (holding one's foot or arm submerged in ice water) instead of a specific emotion.[10] Again, the averaged response patterns for the fear, anger, and pain groups were significantly different from each other. But the response patterns of individual subjects were extremely variable. Guessing which subjects had been angry and which frightened based on their reaction patterns would have yielded a high proportion of mistakes, even though fear and anger seem subjectively quite different and in spite of the very elaborate set of measurements that were obtained in these expensive laboratory investigations. Jones may tend to show Pattern A both when he is angry and when he's afraid, Smith may show Pattern F in both cases, and Fisbee still a different pattern. This variability in autonomic response to the same stimulus has been extensively studied by psychophysiologists under the name *autonomic response-stereotypy*.[11]

More than 100 years ago, William James suggested that emotion consists of the perception of bodily changes, that we are afraid because we tremble, rather than the other way around.[12] If the Jamesian theory were literally correct, then Ax and Schachter should have found distinctive fear and anger patterns in each of their subjects. Jones, Smith, and Fisbee could not have been experiencing the same emotions, no matter what they may have reported, since their bodily reactions were so different. But James's theory has been substantially refuted by Walter Cannon, among others.[13] One of Cannon's arguments cited the work of Marañon, who had studied the psychological effect of stimulating visceral reactions by direct injection of adrenaline.[14] Adrenaline produces the same physiological response pattern that Ax and Schachter found to be associated, on the average, with fear. But most of Marañon's subjects experienced only "as if" emotions: "I feel as if I ought to be afraid but I'm not." Clearly, the essence of emotional response occurs in the brain, not in the gut, although normal emotional experience involves both an essential and, presumably, distinctive brain state together with the perception of peripheral autonomic arousal. A few of Marañon's subjects found that their initial "cold emotion" suddenly warmed up into a genuine feeling of fear or panic. Any stimulus commonly associated with fear is likely to become a conditioned stimulus for fear; the dry mouth, pounding heart, and trembling hands produced by the adrenaline injection is just such a stimulus and may have such a result. James was a superbly observant psychologist and his theory was accurate as a description of an interesting phenomenon, namely, that trembling—or any other fearful behavior—can actually precipitate a genuine emotion. That is why it is adaptive to simulate self-confidence or assertiveness when fear threatens. By avoiding the expressive behavior one normally shows when afraid, one can often stave off fear itself. Similarly, in a study of the effects of high spinal cord injuries on emotional reactions, Hohmann[15] found that these patients systematically reported a diminution in the frequency or intensity of feelings of both fear and anger. Sexual feelings were also attenuated but experiences of sentiment were increased. Presumably, for some emotions, the perception of peripheral arousal plays such an important role that, when visceral sensation is cut off by injury and the normal brain–gut–brain reverberations are thus prevented, that aspect of emotional experience is reduced. But even persons with high cervical injuries, deprived of all sensation lower than the neck, can experience fear and rage and the whole spectrum of emotions. Emotional feeling derives from the excitation of presently unspecifiable centers or networks in the brain; bodily reactions play an ancillary, facilitating role.

If we could identify these hypothetical emotion centers in the brain, and if we were able somehow to monitor the ebb and flow of their activity, then we might indeed be able to determine just how Fisbee is feeling, whether Fisbee wanted us to know or not. Brain centers that play important roles in the emotional apparatus have been identified, and neuropsychologists have been able to produce intense emotional reactions in experimental animals, or to inhibit normal emotional responses, by electrical stimulation in these areas. But the microscopic neural circuits involved in these mechanisms are a thousand times more complex than present techniques can manipulate or present understanding comprehend. Even with the remarkable new methods of brain imaging that can reveal which brain regions increase in activity in response to different stimuli, no neuroscientist with his million-dollar positron emission tomography (PET) or magnetic resonance imaging (MRI) scanners would undertake to tell us *for sure* what O. J. Simpson feels when we ask him that "Did you do it?" question.

If we must forego these expensive devices and measure only his peripheral bodily reactions, interpretation will be hopelessly uncertain. Probably there are distinctive processes in one's brain when one is angry or frightened or, perhaps, guilty; we simply do not know what these processes are or how to measure them. At the periphery, however, all the evidence indicates that different reactions may accompany the same emotion in different people or even in the same person on different occasions. The only aspect of the psychological state that we can measure with any confidence from bodily reactions is unspecified arousal. We may be able to say with some assurance from the polygraph record that this subject was more aroused by Stimulus A than he was by Stimulus B. But we cannot identify the quality of that arousal, whether Stimulus A made him angry or scared or filled with proud delight. Peripheral responses that the polygraph can measure—of muscle tension and voice changes, of heart rate and vasoconstriction, of blood pressure and respiration, of pupil size and palmar sweating—will never provide a basis for emotion detection in this qualitative sense.

We have considered the feasibility of an emotion detector because we had speculated that the feeling of guilt might be the specific lie response, which, it appears, we must be able to identify if we are to develop an actual Truth Verifier, an instrument for detecting lying. It appears that a polygraphic guilt detector is not and will not be feasible. Moreover, it is easy to see that the feeling of guilt could not constitute a specific lie response, anyway. People can and do lie without feeling guilty, some people much of the time, most people some of the time. My wife would have gladly lied, for instance, to keep her sons from being drafted to fight someone else's war and she never would have felt the slightest twinge of guilt. Moreover, many people have had the experience of feeling guilty even when they were telling the truth, expecting not to be believed for some reason or knowing that their (true) story sounds incredible. Like a blush, the feeling of guilt can invade a susceptible person for no other reason than that he or she fears it might. Perhaps it is just as well that we cannot construct a lie detector based on guilt detection since such a device would be inevitably undependable.

The Current Status of the Specific Lie Response

To say that there is no peripheral, measurable specific lie response implies, as we have seen, that the instrumental lie detector of popular mythology cannot exist. But this is not to say that lie detection is impossible, and later chapters are devoted to discussing how this is attempted and how well it works. I talk about *guilt* detection as well, not the detection of the emotion or feeling of guilt, but rather the detection of guilty knowledge: Does the respondent know about or recognize events or faces or objects of which he claims to be ignorant, knowledge that only the guilty suspect might be expected to have?

But the phantom of the old specific lie response still lurks about, confusing laypersons and professionals alike, and we should try to bury it once and for all with a stake through its heart if we can manage it. The problem is complicated by inconsistencies, both logical and semantic, in the literature of polygraphy. The Control Question Test was developed precisely because no specific lie response exists. An innocent but reactive subject might respond strongly to the relevant or "Did you do it?" questions, so most modern examiners ask control questions, unrelated to the issue under investigation, to provide an index of the subjects' reactivity against which to compare their responses to relevant questions. Yet the inventor of the control question technique, John Reid, in the most recent edition of his widely used textbook on polygraphy, devoted pages of text and illustration to what he calls "deception criteria" or "typical deception responses."[16] For example, shallow breathing, heavy breathing, a speeding up or a slowing down of respiration, a "sigh of relief" after a crucial question are all represented as being "dependable" or "very reliable" criteria of deception. There is no objective evidence to support such claims.

The layperson will find it hard to understand how honorable men, respected in their field, can persist in making claims, wholly unverified by systematic, controlled study, that almost certainly are false. I consider this phenomenon to be neither surprising nor scandalous. An experienced examiner like Reid accumulates years of unsystematic observations, a fertile source of hypothesis and speculation. But, in the vast majority of those thousands of polygraph examinations, he will have had no way of knowing independently and with certainty whether the respondent was lying or not. If "staircase respiration" is found to be associated with a known lie on one occasion and then is observed again with nine other subjects who might have been lying (but are never proved to be), the temptation seems almost irresistible to conclude that this response pattern is a "dependable criterion of deception."

The history of polygraphy is littered with similar claims, forcibly asserted and honestly believed by their proponents, claims that have since been proven to be wrong. Vittorio Benussi claimed to have demonstrated in 1914 that lying was reliably accompanied by a change in the ratio of expiration to inspiration, but no one has taken Benussi's ratios seriously for years. Father W. G. Summers asserted in 1937 that if the "Did you do it?" question is repeated several times, the electrodermal response produced by each question will increase if the subject is lying but will decrease if he is being truthful. Summers reported 100% accuracy in detecting deception with this method on a series of criminal cases. Father Summers was chairman of the psychology department at Fordham University, not only a man of high integrity but a trained scientist. And yet few modern polygraphers have even heard of either Summers's criterion or his Pathometer. W. M. Marston reported in 1915 that lying produces an increase in systolic blood pressure, and 23 years later he described experiments seeming to corroborate these claims with accuracies of 97% to 99%. Ominously, in one of these experiments, Marston also tested both Benussi's and Summers's claims and rejected them. Subsequent studies by others similarly have refuted all three claimants.

Marston himself remarked in 1938 that "some of the most amusing lies on record have been told in connection with the Lie Detector itself." I would substitute "untruths" for "lies." It seems most unlikely that all or any of these competing claims, including Marston's own, involved deliberate deception. They were the product instead of overenthusiasm, a lack of adequate criteria, and generally inadequate measurement techniques. Consider just one example. For more than 50 years polygraphers believed that they could assess blood pressure changes from the "cardio" tracing of the polygram, the tracing drawn by the pen connected to the arm cuff. Some workers in fact believed that the upper and lower margins of the tracing, respectively, changed with variations in systolic and diastolic blood pressure. In 1977, Geddes and Newberg demonstrated that the behavior of the "cardio" tracing depends entirely on whether the subject's mean blood pressure is higher or lower than the pressure to which the examiner happens to have pumped up the cuff.[17] Thus, the type of response that examiners have universally been interpreting as a blood pressure increase may actually be produced by a decrease in pressure and vice versa.

In addition to lack of independent criteria ("Was this subject really lying or not?") and inadequate measurement techniques ("Was this an increase in blood pressure or a decrease?"), mythology about alleged specific lie responses is encouraged by the sheer complexity of the polygraph recordings on which, to borrow I. Younger's graphic phrase, "the needles move as subtly as Raskolnikov's soul."[18] Those four squiggly lines can change in so many ways that it is almost always possible to find some change or pattern that, on a particular chart or set of charts, seems to have some particular significance.

In 1978, for example, Raskin and Hare reported results of a mock crime experiment in which it could be certainly established which answers were lies and which were truthful.[19] These authors were able to look for literally dozens of physiological changes that might have been associated with known lies. They do not report success with the Benussi or the Summers or the Marston stigmata but instead claim some new ones, changes involving "abdominal respiration baseline" and the amount of heart rate deceleration following the initial acceleration usually produced by a relevant question. Because of faulty statistical analysis, Raskin and Hare cannot even properly assert that their liars and truth-tellers differed on the average with respect to these complex patterns. What their study really shows, like so many others, is that no pattern of physiological response is unique to lying.

I can only agree with the comment by R. J. Weir, Jr., a past president of the American Polygraph Association and director of polygraph operations for "a federal agency" (a "spook" phrase that usually denotes the CIA) for many years:

> I have even heard experienced examiners get mousetrapped into a discussion as to whether there is some mysterious difference between the reactions created by lies and those from strong emotions such as fear or anger. All I know is that I know of no way to make this distinction merely from the chart patterns.[20]

A polygraph examiner who asserts that a respondent "showed deception" or "gave a deception response" on a particular question is making either a misstatement or a false statement. He may be entitled to say, based on the charts, that the respondent made a larger response to one question than to another, and the examiner may proceed to draw some inference based on this difference. But the myth of the specific lie response should be laid at last, permanently, to rest.

Chapter 5

EVALUATING
THE EVIDENCE

There are three kinds of lies; lies, damn lies, and statistics.

—Benjamin Disraeli

As psychologists we do not trust our memories, and have no recourse except to record our predictions at the time, allow them to accumulate, and ultimately tally them up. We do not do this because we have a scientific obsession, but simply because we know there is a difference between veridical knowledge and purported knowledge, between knowledge that brings its credentials with it and that which does not. After we tally our predictions, the question of success (hits) must be decided upon. If we remember that we are psychologists, this must be done, either by some objective criterion, or by some disinterested judge who is not aware of the predictions.

—Paul E. Meehl[1]

How dependable is the lie detector, then? Here is a straightforward question for which there is no simple answer. Since "dependable" is vague and the "lie detector" does not exist, I must start by rewording the question. There are several different types of polygraphic examination, each based on different assumptions; one cannot assume that all these types will have the same degree of accuracy. To provide some initial perspective, remember that the purpose of a polygraph examination is to diagnose the individual respondent as deceptive or truthful with greater accuracy than one could achieve without the examination. I can classify subjects as truthful or deceptive and be correct half the time merely by flipping a coin; the *chance accuracy* of this type of dichotomous classification is 50%—if 50% of the subjects are actually lying. If my subjects are all defendants who have been brought to trial on criminal charges, and if the statistics show that 80% of this group, on the average, are in fact guilty, then I could attain 80% accuracy just by classifying everyone as deceptive. For a test to be useful, its accuracy must obviously be higher than one can achieve by chance, and it should usually be higher than the base rate of the more frequent classification in the group tested.

Finally, we must ponder the deeper question of how accurate a lie test ought to be for particular applications. If one is hiring policemen or CIA operatives, then it might be thought that any additional clues, any improvement over chance at all might be worthwhile. These are sensitive positions in which the wrong person can do great mischief, and it may be in the public interest to use a screening procedure that reduces the number of undesirable candidates hired, even if this means also excluding a large number of perfectly acceptable people, wrongly called deceptive by the test. As we shall see later, however, there is reason to believe that many honorable people, the very sort of "straight arrows" we should like to see in these sensitive positions, are especially vulnerable to failing and being eliminated by these screening tests. Moreover, cases like that of Aldrich Ames indicate that *false-negative* errors (classifying a liar as truthful) not only occur but do great harm when reasonable suspicions are quieted by unjustified faith in the polygraph. In the United Nations fantasy that we considered earlier, it would be disastrous to settle for even 90% accuracy in the Truth Verifier. If so much weight is placed on the test result that one makes less effort to seek other information or is lulled into a feeling of great confidence in the result, then one should make sure that the test result is very dependable indeed.

The Limitations of Expert Opinion

In the standard textbook of polygraphic interrogation, Reid and Inbau assert,

Our actual case experiences over the years have involved the poly-graph examination (either personally or under our direct supervision) of over 100,000 persons suspected or accused of criminal offenses or involved in personnel investigations initiated by their employers. On the basis of that experience, we are confident that the technique, when properly applied by a trained, competent examiner, is very accurate in its indications. The percentage of known errors with the technique used in the laboratories of John E. Reid and Associates is less than 1 percent.[2]

Another highly regarded polygrapher of wide experience, R. O. Arther, similarly claims an accuracy of 99%.[3] In 1939, the chairman of the psychology department at Fordham University, the Reverend Walter G. Summers, claimed 100% accuracy on more than 200 criminal cases.[4] Testifying before a committee of the Minnesota state legislature in 1975, a polygrapher from Texas stated that he had given more than 20,000 lie tests in his career and had "never been shown to have made a mistake." David Raskin, a former professor of psychology and primarily responsible for the marriage of the polygraph to the personal computer, reported in 1983 to a federal judge in California that the computer indicated a probability of *100%*(!) that John DeLorean was truthful in denying his guilt on a drug charge. Paul Minor, then head of the FBI's polygraph unit, subsequently tested DeLorean and found him to be deceptive. Mr. Minor recently asserted on national television that the lie detector's error rate is only "one to two percent."[5]

These are not selected examples. Nearly every experienced polygraphic examiner who has recorded an opinion about the accuracy of tests he has himself administered has chosen an estimate in this range, where 95% is "conservative" and 99% is perhaps typical. And most of these polygraphers are honorable people; it would be absurd to accuse all of them of venal misrepresentation. In many seemingly parallel situations, both in the courtroom and in everyday life, the opinions of such experts, based on their long experience, are taken very seriously.

One must realize, first, that someone who has devoted a career to lie detection, who has given thousands of tests the results of which have seriously affected for good or ill the lives of many people, must inevitably be strongly motivated to believe that these tests have been accurate. Experienced polygraphers would be less than human if they were not quicker to perceive positive than negative evidence of the value of their work. Second, the utility of polygraph testing does not depend solely on the accuracy of the lie test. The polygraph examination acts as a powerful inducer of admissions or confessions and, because of the mystique of the procedure, would do so even if the polygraph were just a stage prop. Examiners who are frequently able to elicit admissions of misconduct or, in criminal cases, admissions of guilt may therefore feel that they control a powerful technique—and "powerful" is easily transmuted into "valid." Moreover, like everyone else, polygraphers are more inclined to remember the good cases than the bad ones and to have a clearer recollection of those instances where their efforts solved some mystery than the ones where they remained in doubt.

These considerations are especially important because, *in the vast majority of examinations, polygraphers never know if they were right or wrong*. In criminal cases, many crimes are never solved, most suspects never go to trial. How then do we account for the claims of 95% and 100% accuracy? We must attribute them to the inevitable distortion that results when true believers attempt to evaluate the soundness of their own beliefs using "noisy" and inadequate data.[6]

How Polygraph-Induced Confessions Mislead Polygraphers

It is standard practice for police polygraphers to interrogate a suspect who has failed the lie test. They tell him that the impartial, scientific polygraph has demonstrated his guilt, that no one now will believe his denials, and that his most sensible action at this point would be to confess and try to negotiate the best terms that he can. This is strong stuff, and what the examiner says to the suspect is especially convincing and effective because the examiner genuinely believes it himself. Police experience in the United States suggests that as many as 40% of interrogated suspects do actually confess in this situation. And these confessions provide virtually the *only* feedback of "ground truth" or criterion data that is ever available to a polygraph examiner.

If a suspect *passes* the polygraph test, he will not be interrogated because the examiner firmly believes he has been truthful. Suspects who are not interrogated do not confess, of course. This means that the only criterion data that are systematically sought—and occasionally obtained— are confessions by people who have failed the polygraph, confessions that are *guaranteed* to corroborate the tests that elicited those confessions. The examiner almost never discovers that a suspect he diagnosed as truthful was in fact deceptive, because that bad news is excluded by his dependence on immediate confessions for verification. Moreover, these periodic confessions provide a diet of consistently good news that confirms the examiner's belief that the lie test is nearly infallible. Note that the examiner's client or employer also hears about these same confessions and is also protected from learning about most of the polygrapher's mistakes.

Sometimes a confession can verify, not only the test that produced it, but also a previous test that resulted in a diagnosis of truthful. This can happen when there is more than one suspect in the same crime, so that the confession of one person reveals that the alternative suspect must be innocent. Once again, however, the examiner is usually protected from learning when he has made an error. If the suspect who was tested first is diagnosed as deceptive, then the alternative suspect—who might be the guilty one—is seldom tested at all because the examiner believes that the case was solved by that first failed test. This means that only rarely does a confession prove that someone who has already failed his test is actually innocent.

Therefore, when a confession allows us to evaluate the accuracy of the test given to a person cleared by that confession, then once again the news will almost always be good news; that innocent suspect will be found to have passed his lie test, because if the first suspect had not passed the test, the second person would not have been tested and would not have confessed.[7]

A Real-Life Illustration   Here is an example of how this process works in real life. In *Polygraph*, the trade journal of lie detection, a police examiner named Murray reported on a series of 552 lie tests that he administered to criminal suspects.[8] He diagnosed 239, or about 43%, of the total group to be deceptive (see Table 5.1). Murray proceeded to interrogate these 239 people and obtained confessions from 105, or nearly half of them. Murray assumed that these verified-deceptive lie tests were representative of all 239 failed tests, but of course that was a mistake. *All* of the 105 lie tests that produced these 105 confessions were *necessarily* verified as accurate.

These confessions also cleared 18 of the 313 suspects who had previously been tested and classified as truthful. Once again, Murray assumed that these 18 verified tests were representative of all 313 lie tests that were passed, but once again, this was an invalid inference. As we have seen, once a prior suspect has failed his lie test, alternative suspects in the same case are seldom tested at all; therefore, these 18 successes were also the

Table 5.1. The Murray (1989) Study of 552 Criminal Suspects

Lie test results	Correct	Errors	Unknown
239 (43%) fail. All interrogated. 105 (44%) confess.	105 (44%)	3 (1%)	131 (55%)
313 (57%) pass. No interrogation. None confess.	18 (6%)	0	295 (94%)
Totals	123 (22%)	3 (0.5%)	426 (77%)

almost inevitable consequences of reliance on confessions as the only criterion.

Mr. Murray does report three cases in which he discovered that he had made a false-positive error. These were cases in which there were two possible suspects; the first person tested had failed his test but Murray was suspicious of the result and broke his own rule by going on to test the other suspect. In each of the three cases, the second person tested also failed, was interrogated, and confessed.

In one case, for example, the first suspect was an old lag, an habitual criminal whom Murray had happened to test on two previous occasions for other crimes. Both prior times, this man had failed the test, had been interrogated, and confessed. On this third occasion, he again failed the lie test but this time he continued to maintain his innocence. For this reason, Murray tested the other suspect, obtained a confession, and discovered that the habitual criminal was innocent of this third crime and that his third lie test had been in error. Murray concluded, however, that he had made only these three errors in 552 lie tests, an accuracy of 99.4%.

It is important to see that Murray would have obtained much the same apparent confirmation if, instead of the polygraph, he had used for his lie test just the flip of a coin! About half of the coins would have come up heads, indicating deception (43% of them, if we use Murray's statistics.) About half of these purportedly deceptive subjects would have been guilty and many of the guilty ones, perhaps 105 of them, would have confessed following interrogation. Murray would be protected from learning, however, that most of the 134 persons who also came up heads but refused to confess were in fact innocent.

Once again, a few of these confessions would have cleared some—perhaps 18—of the 313 suspects whose tests came up tails for truthful. Murray would never learn, however, that about half of the 313 people who passed his coin-flip lie test were in fact guilty. As this analysis reveals, the findings that Murray was so proud to report are entirely compatible with the assumption that his lie detector diagnoses were only randomly related to the truth! Polygraphers are not scientists or statisticians. When 20% to 25% of their tests are verified by confession and they stumble upon errors as seldom as 3 times in 552, who can blame them for thinking they are nearly infallible? They are victims of their own deceptive art.

That examiner who testified before the Minnesota legislature said that in 20,000 cases he had never been proved to be wrong, but he did not reveal how often he had been proved to be right. One can only assume that until someone somehow proves him wrong he will continue honestly to think that he is an infallible lie detector. One might further suspect that if that one mistake were proved, he would continue to claim, honestly but incorrectly, that his batting average was (20,000/20,001) or much better than 99.9%. The chief polygrapher for a federal agency, a past president of the American Polygraph Association, recently wrote, "I have never seen an authenticated case of a 'false-positive' presented yet"—that is, a truthful subject erroneously diagnosed as deceptive. Here is a breathtaking example of what the Church might call invincible ignorance; here are eyes that will not see.

The conclusion seems inescapable that the opinion of the professional examiner as to the accuracy of the polygraph test, based on his clinical experience, however extensive, is of negligible probative value. He is in the position of the astrologer who believes in what he is doing but almost never has a chance to actually test whether his predictions are correct or not. But, like astrological predictions—or psychological test results or medical diagnoses—the accuracy of polygraph diagnoses *can* be assessed by objective, controlled scientific study. A number of such studies have been reported, ranging in quality from reasonably good to plainly worthless. Surveys of this literature have tended to ignore these wide differences in the quality and relevance of this material. Abrams, for example, tabulates a dozen such studies spanning 56 years, some using the relevant/irrelevant test and others the Control Question Test, studies based on as few as 11 cases or as many as 157, studies using clinical evaluation and others using independent chart evaluation—and then Abrams simply averages the percent correct claimed in each study.[9] I shall try to avoid such a meaningless exercise.

Lykken's Law

C. Northcote Parkinson cannot have been the first person to notice that bureaucratic activity expands to fill whatever time or space has been allocated to it. But he codified this principle, now known as Parkinson's Law.[10] L. J. Peter was not the first to realize that competent people keep getting promoted until they reach the level at which their abilities are overtaxed. When succinctly stated as the Peter Principle, however—"administrators rise to the level of their incompetence"—this insight became much more generally available.[11] Although Peter and Parkinson may not have been the first explorers in their respective areas, they made the first maps and thereby performed a public service.

The territory that I should like to map is one in which people try to make evaluations or predictions based on a mixture of evidence, no piece of which is conclusive by itself. A graduate school committee deciding which student applicants should be admitted, a jury evaluating the evidence against a defendant, an employer deciding whom to hire—in all these situations people are confronted with a collection of relevant information, some of it pointing in one direction and some in the other. The task is to attach weights to each bit of evidence—How important is it? How much do I trust it?—and then to combine them in arriving at a final decision. The graduate student's application, for example, will include letters of reference, the student's essay explaining why she wants to enter the program and what she thinks her strengths are, the list of the undergraduate courses she has taken, copies of papers she has written, her grade point average, and her aptitude test score. Which of these pieces of evidence will have the most weight in the final decision? Almost invariably, the students admitted will turn out to have the highest test scores and the best grade point averages.

Grades of A in Beginning Volleyball or Intermediate Basket Weaving may not tell us as much about a student's potential as do grades of A or even B in Calculus, Theoretical Physics, or Microbiology. Yet the student who elects tough courses and does well in them may have to yield her place to one who achieves straight A's by taking easier courses. A B+ average from Harvard may promise more than an A average from Chickesaw State. Yet, in the press of decision making, the committee's eyes are drawn to that unblemished record. For decades, my own university department has used a high-level aptitude test, the Miller Analogies Test, in selecting graduate students. I once did a study of the way in which more than 150 of our students actually performed in graduate school, in relation to their MAT scores. There was no discernible relationship. Those scoring above the 90th percentile on the MAT were just as likely to fail or drop out as they were to be rated "outstanding." But we continued to use the Miller. Grade averages and aptitude test scores are so simple and unambiguous. They make it possible to rank people objectively and without argument.

Uncertainty is painful to the decision maker. Complicated evidence can only be evaluated subjectively and subjectivity leads to doubt and disagreement. One longs for some straightforward, definitive datum that will resolve the conflict and impel a conclusion. This longing not infrequently leads one to invest any simple, quantitative, or otherwise specific bit of evidence with a greater weight than it deserves, with a predictive power that it does not really possess. In decision making, the objective dominates the subjective, the simple squeezes out the complicated, the quantitative gets more weight than the nonmetrical, and dichotomous (yes/no, pass/fail) evidence supersedes the many-valued. This is Lykken's Law.

One reason why the lie detector has become so important in American decision making can be understood in terms of this principle. However subjective the process may be of arriving at a lie test diagnosis, the diagnosis itself is objective from the point of view of the employer or the jury; the job applicant or the defendant either passed or failed. Nothing could be simpler than that. This policeman may have a spotless record; he may have gone out of his way to return money that accidentally came into his possession and that he could have kept. But he flunked the lie detector test and so we conclude that he accepted a bribe to smuggle a weapon into the jail. This store manager is known in the community as a devout family man who lives modestly with no sign of affluence. But he failed the lie test so we believe that he has been stealing thousands from the store. This rape victim relates a harrowing account of being threatened, abused, and violated. But she could not pass the lie test, so we conclude that the sex act was consensual and refuse to prosecute the man that she accuses. The results of a polygraph examination are almost always simple and specific. But we must not confuse simplicity with validity. The lie test diagnosis may be unambiguous; the important question is whether it is correct.

Reliability versus Validity

The layperson is inclined to use the terms "validity" and "reliability" interchangeably when referring to the accuracy of a test. But there are two aspects to test accuracy that it is important to distinguish and for which it is useful to have separate labels. In standard psychometric parlance, the *reliability* of a test is the consistency with which the test measures whatever it is that it measures. The *validity* of a test is the extent to which the test measures that which it is claimed to measure, that is, its accuracy. On one hand, a test can be highly reliable but have low validity; on the other hand, a test with low reliability cannot have high validity.

Reliability   Many psychological tests, IQ tests for example, are intended to measure the amount of some trait possessed by the respondent. Tests used in polygraphic interrogation, however, are usually designed to classify respondents into one of two categories, deceptive versus truthful, or guilty versus innocent. The reliability of a lie test—the consistency with which it classifies people—might be estimated by having the same group of respondents tested twice, by two different examiners. If the test were perfectly reliable, the two examiners' decisions should agree 100% of the time. If the reliability of the test were zero, then the percent of agreement will not be zero but, rather, about 50%. For example, suppose the "test" used by each of the examiners was just the flip of a coin—heads you're truthful, tails you're lying. Such a "test" would have no reliability at all; how you were classified the second time around would have no relationship to whether you passed or failed on the first testing.

This situation is illustrated in Table 5.2. We would expect each examiner (each "testing") to classify half the people as deceptive and half as truthful. But only 50% of the people called deceptive on the first test would be expected to be called deceptive on the second test also. The total agreement expected between the two examiners is obtained by adding the number of persons called deceptive by both tests, plus the number called truthful by both. For this case of a wholly unreliable test, the expected rate of total agreement is 50%.

But this chance level of agreement may differ from 50% if the two examiners "fail" more or less than half the group. Suppose Examiner A thinks that about 90% of the suspects are probably guilty, so he simply passes every tenth person he tests and fails all the rest. And suppose

Table 5.2. A Hypothetical Reliability Study
Using a Coin Flip to Classify People
as Deceptive or Truthful

	EXAMINER B		
EXAMINER A	Deceptive	Truthful	Total
Deceptive	25	25	50
Truthful	25	25	50
Total	50	50	100

Note. It is assumed that Examiner A "tests" the 100 subjects first, then Examiner B tests them again. The total agreement between the two "tests" will be 50% (on the average).

Examiner B, slightly less cynical, passes every fifth person tested, thus failing 80% compared to 90% for A. If B tests the people in a different sequence than does A, then we can say again that these "tests" have a reliability of zero. Nevertheless, these two (wholly imaginary!) examiners, just by summarily flunking 80% or 90% of all people they test, choosing their victims arbitrarily, would nonetheless agree on 74% of their scorings! This result is illustrated in Table 5.3, which assumes that 100 people are tested altogether, first by Examiner A who fails 90 of them at random, and then by B who fails 80, also at random. Then, of the 90 people failed by A, we would expect B to fail 80%, or 72 people. Of the 10 people called truthful by A, B should call 8 deceptive and 2 truthful. Hence, A and B should agree on 72 + 2, or 74, of the 100 cases, just by chance alone.

What this all means is that *the total percentage of agreement is a poor way of assessing the reliability of a polygraph test*. It is also a poor way of measur-

Table 5.3. Expected Results
of a "Reliability Study"

	EXAMINER B		
EXAMINER A	Deceptive	Truthful	Total
Deceptive	72	18	90
Truthful	8	2	10
Total	80	20	100

Note. In this study Examiner A randomly calls 90% of a group of 100 subjects deceptive and Examiner B randomly calls 80% of the same group deceptive also.

ing the validity of this type of test, as we shall see later. The correct way to find the test reliability from data like those in Table 5.2 or 5.3 is as follows: (1) Find A's percentage agreement with B on those two cases B called truthful; in Table 5.3 this would be 100(2/20), or 10%. (2) Find A's agreement with B on those cases called deceptive by B; in Table 5.3, this would be 100(72/80), or 90%. (3) Average these two percentages; in Table 5.3, this would give (10 + 90)/2, or 50% agreement.

With this method, we shall always be able to say that 50% agreement represents chance expectation or zero reliability. Perfect reliability, of course, will always be indicated by 100% agreement. The reliability of actual tests will fall between 50% and 100%. The method works even in extreme cases like the one illustrated in Table 5.4. Here, two examiners both test 100 residents of a dormitory to discover which one is guilty of having stolen some money. Under such circumstances, each examiner will naturally expect to find only one culprit, and for the sake of this illustration, we shall assume that Examiner B identifies a different culprit than Examiner A chooses. Notice that in this case, even though the two examiners completely disagree about who the guilty party was, their total agreement, figured in the usual way, would be 98 + 0, or 98%! Calculated in the correct manner, however, their agreement (the test reliability) is 49.5%, close enough to 50% for our purposes.

The examples so far have been chosen to show what the data might look like with totally unreliable tests and should not be taken to suggest that real polygraph tests behave in this manner. To redress this imbalance,

Table 5.4. Examiners A and B Each Identify
a Different One of the 100 Residents
of a Dormitory as the Possible Thief

| | EXAMINER B | | |
EXAMINER A	Deceptive	Truthful	Total
Deceptive	0	1	1
Truthful	1	98	99
Total	1	99	100

Note. Total agreement is 98%, but the corrected average agreement is obtained by averaging 99%, the agreement on the cases B called truthful, with 0%, the agreement on the cases B called deceptive, getting 49.5%, which indicates a reliability of zero (fictitious data).

Table 5.5. Actual Lie Test Reliability Data

| | EXAMINER B | | |
EXAMINER A	Deceptive	Truthful	Total
Deceptive	36	8	44
Truthful	8	60	68
Total	44	68	112
Percent agreement	82%	88%	85%

Note. Calculated from data reported by Horvath on 112 criminal suspects whose charts were read independently by ten examiners (F. S. Horvath. The effect of selected variables on interpretation of polygraph records, *Journal of Applied Psychology*, 1977, *62*, 127–136). Data shown represent the average of all possible pairings of the 10 independent examiners.

some actual lie test reliability data are given in Table 5.5 showing 85% agreement, much better than chance but still leaving 15% of subjects diagnosed as deceptive by one examiner and as truthful by another.

A few polygraphers currently are using computerized polygraphs in which the electrical signals representing the subject's physiological reactions are measured and scored automatically by a computer algorithm. Computer scoring is ideal from the viewpoint of reliability because the computer (almost) always gives the same result. Here we must emphasize again, however, that reliability and validity are two different things. Properly programmed, a computer can give an astrological prediction that is perfectly reliable—but has only chance validity.

Validity   Whereas the reliability of a lie test is measured by the agreement between repeated tests, or, most commonly, between independent scorings of the same tests, the *validity* of a lie test is measured by the agreement between the results of the test and "ground truth"—which respondents were in fact lying and which truthful. Therefore, to assess the validity of any form of polygraph test, it is necessary to obtain a *criterion measure* against which to compare the test results. The criterion is easily obtained in a laboratory experiment where one can decide in advance which subjects will lie and which will tell the truth on the test. In field situations, for obvious reasons, it is usually much harder to obtain dependable criterion measures and one may have to make do with a criterion that is less than perfect. In a series of criminal cases, for example, we might say that those subjects who are ultimately found guilty by a court must have been lying during the earlier polygraph test, while those who are exonerated by a "not guilty" verdict or by the conviction of another person might be said to have been truthful when earlier tested. But judicial decisions are wrong at least some of the time and, thus, must be regarded as an imperfect criterion. The strict rules of evidence observed by our courts sometimes allow a manifestly guilty defendant to escape conviction. Therefore, a possible improvement over using formal adjudication as a criterion is to use instead the consensus of a panel of legal experts who review the case facts and, uninhibited by the limitations on admissibility that must constrain the courts, decide whether a criminal suspect was probably innocent or guilty or whether the facts are insufficient to permit a judgment to be made.

Finally, confessions can be used as a criterion. Suspects who have been tested may subsequently confess their guilt, showing that they had lied on the test, whereas others may subsequently be cleared by the confession of another person. As we have seen, a failed lie test stimulates determined interrogation, and the knowledge that he has failed the test helps persuade the guilty suspect to abandon further efforts to deceive. This means that studies using confession as the criterion will tend to exclude that type of guilty suspect who produces a false-negative lie test; having "beaten" the lie test, such suspects are not interrogated by the polygrapher and are less likely ultimately to confess. Such studies, therefore, will tend to be biased in the direction of showing inflated validities for the criterion-guilty group.

In spite of these difficulties, a reasonably trustworthy criterion must be obtained if we are to estimate the validity of any form of polygraph test, and it is essential that the accuracy of so important a decision-making procedure be established empirically. For many years, the validity of the lie test has been largely assumed: "If he is so disturbed by the question, then he must be lying." More recently, commonsense arguments have been vigorously advanced against the lie detector: "You would be disturbed by these accusatory questions, too, if you were the innocent accused." Implausible assumptions do sometimes turn out to be correct, however, and plausible arguments are sometimes wrong. The final answer must be empirical. In a properly conducted validity study, using an acceptable criterion of ground truth, in what proportion of cases does the given form of polygraph test agree with the criterion?

Percentage agreement with the criterion is not an adequate measure

Table 5.6. Hypothetical Example Illustrating
How a Wholly Invalid Test Might Produce as High
as (81 + 1) = 82% Agreement with the Truth

| CIA TEST RESULT | FACTS KNOWN TO THE OMNISCIENT | | |
	Deceptive	Truthful	Total
Deceptive	1	9	10
Truthful	9	81	90
Total	10	90	100
Percent agreement	10%	90%	50%

of test validity, however, since even a totally invalid test might agree with the criterion in a high proportion of cases. This important point is illustrated in Table 5.6, which is based on the following imaginary circumstances: Suppose that, before the Berlin Wall came down, the Soviets had managed to plant 11 moles among the 101 top executives of the American CIA. Suppose also that, after the Aldrich Ames debacle, all the remaining 100 were scheduled for repeat lie tests. Because of their embarrassment about Ames, suppose that the CIA polygraphers tighten the screws of their procedures, failing 10% of the total group, but, unknown to them, the actual validity of their tests is no better than chance. As shown in Table 5.6, under the peculiar circumstances hypothesized, these invalid tests would none the less turn out to be correct 82% of the time! By passing 90% of the respondents at random, they would be likely to pass 81 of the 90 loyal executives. By failing 10% at random, 1 of the 10 planted moles would be expected to be "identified." Of course, only the Omniscient would know of that misleading 82% result and He would also know how misleading it was. The CIA examiners, however, would congratulate themselves on having caught that additional mole and continue to believe that their tests were well-nigh infallible (until, say, Harold Nicholson was uncovered).

The problem illustrated by this example is similar to the one we considered in searching for a method of expressing lie test reliability. The solution is similar also. Instead of using total agreement between test and criterion, which may be influenced by factors unrelated to test accuracy, we simply compute the test's performance with the criterion-truthful and the criterion-deceptive subjects separately and then average these two percentages. For the fictitious data in Table 5.6, the random "lie test" achieved an apparent accuracy of 90% on the criterion-truthful subjects but was correct on only 10% of the criterion-deceptive respondents; these two numbers yield a mean accuracy of 50%, which, as the example illustrates, is the accuracy expected on the basis of chance alone.

For a more realistic example, let us consider data from a validity study by Drs. G. Barland and D. Raskin, reported in 1977 to the Law Enforcement Assistance Administration. These data are shown in Table 5.7.[13] For convenience, the data have been expressed as percentages to facilitate comparison with Table 5.6.

Notice that the total agreement between lie test and criterion is equal to (39 + 5)/51, or 86%, but this does not tell us the validity of the lie test used. It happened in this set of data that most (40/51 = 78%) of the persons tested were criterion-guilty. It also happened that Raskin scored most (45/51 = 88%) of the polygraph charts as indicating deception. Therefore, the test would have classified 71% of the subjects correctly by random allocation of the 45 "fails" and 6 "passes." To get a realistic assessment of the test validity, we must consider its accuracy on the guilty and innocent subjects separately. On the criterion-guilty, the lie test accomplished 97% correct classification; on the criterion-innocents, it was correct only 45% of the time; the true validity, therefore, is (97 + 45)/2, or 71%, which (coincidentally?) equals exactly the 71% correct classifications that would be expected in this case just by chance alone.

A fairly common application of the polygraph is the situation in which a crime has been committed and numerous individuals are poten-

Table 5.7. Validity Data from Barland and Raskin

RASKIN'S DIAGNOSIS	JUDGE'S ASSESSMENT OF GUILT		Total
	Guilty	Innocent	
Deceptive	39	6	45
Truthful	1	5	6
Total	40	11	51
Percent agreement	97%	45%	44/51 = 86%

Note. Cases have been omitted for which either the lie test or the criterion judgment was inconclusive. The actual number of cases (grand total) was 51; the figures have been converted to percentages for easier comparison with Table 5.6. Average agreement on guiltys and innocents = (97% + 45%)/2 = 71%.

tial suspects, of whom only one is plausibly guilty. In such instances, the examiner will be most reluctant to classify more than one subject as deceptive. Thus, for example, investigating a theft in a dormitory with 81 residents, psychologists Bitterman and Marcuse found in 1947 that 7 residents produced apparently deceptive records on the first testing using the relevant/irrelevant format.[14] After one to five retests, all 7 gave apparently truthful records, so these investigators were led to conclude that none of the 81 residents were guilty of theft. If only 1 of the 81 persons tested had produced a deceptive result on the first testing, one supposes that there would have been less insistence on repeated retesting.

These single-culprit situations, while not without interest, are especially susceptible to misinterpretation. One real-life example is illustrated in Table 5.8. In 1975, a supermarket in a South Carolina town experienced extensive losses. The parent company sent a team of two commercial polygraphers to test each of the store's 52 employees. As a result of this testing, the assistant store manager was identified as being responsible for the losses and he was discharged. The company subsequently published an apology, acknowledging the lack of any evidence of this man's misconduct, so we can reasonably assume that the lie test that identified him as deceptive was inaccurate. Assuming that the cited losses were real, it is reasonable to suppose that some other employee was responsible; if so, we might conclude that one of the other lie tests, which classified the true culprit as truthful, was also in error. This result is illustrated in Table 5.8, where we can see that even though the lie test wrongly stigmatized an

Table 5.8. Single-Culprit Scenario

LIE TEST RESULT	CRITERION		
	Deceptive	Truthful	Total
Deceptive	0	1	1
Truthful	1*	50	51
Total	1	51	52
Percent agreement	0%	98%	50/52 = 96%

Note. Fifty-two supermarket employees are interrogated, one wrongly diagnosed as deceptive. It is assumed that another employee, classified as truthful, was actually guilty. Although total agreement equals 96%, the mean accuracy on truthful and deceptive subjects considered separately equals (98 + 0)/2, or 49%.
*The inclusion of one false-negative is merely an assumption; none—or several—of the persons classified as truthful may have been deceptive.

innocent person *and* failed to identify the actual thief, the total accuracy of the procedure could nonetheless be evaluated as 48/50, or 96%!

Correctly evaluated, however, as the average of 98% accuracy on the innocent respondents and 0% accuracy on the guilty, we obtain a chance accuracy figure of 49%. This same example illustrates one weakness of the method advocated for computing average accuracies as estimates of true validity. If we assume that all 52 of these employees were in fact innocent of wrongdoing, then there would have been no criterion-deceptive cases on which to base an average accuracy. As is generally true of statistical procedures, no rule is foolproof.

Assessing Lie Detector Validity

Two types of studies have been used to determine whether polygraph tests work. In laboratory or *analog* studies, volunteer subjects, often college students, commit or do not commit mock crimes and are then subjected to polygraph tests. These tests are scored and the percentage of "guilty" subjects who are detected as deceptive (the *true-positive* percentage) and the proportion of "innocent" subjects detected as truthful (the *true-negative* rate) are determined. The great advantage of the analog method is that one has certain knowledge of ground truth, of which subjects are lying and which are not.

Laboratory studies, however, have serious disadvantages for predicting lie detector accuracy in real-life criminal investigations.

1. The volunteer subjects are unlikely to be representative of criminal suspects in real life.
2. The volunteers may not feel a lifelike concern about mock crimes that they have been *instructed* to commit and about telling lies they are *instructed* to tell.
3. Compared to criminal suspects, who know they may be in real trouble should they fail the lie test, volunteers are unlikely to be as apprehensive about being tested with respect to mock crimes for which they will not be punished, irrespective of the test's outcome.
4. The administration of the polygraph tests tends not to resemble the procedures followed in real life. For example, unlike real-life tests, which are most often conducted well after the crime took place, laboratory subjects are typically tested immediately after they com-

mit the mock crime. Moreover, in laboratory research, to make the study scientifically acceptable, there is an attempt to standardize the procedure (e.g., all subjects are asked identical questions), a factor that distinguishes these from real-life tests.

The many problems with laboratory studies indicate that their results are not generalizable to real-life applications of polygraph testing. As we shall see later, the vast majority of scientists whose opinions were surveyed about this issue agreed with this conclusion.

Another approach to evaluating the validity of polygraph tests is to rely on data collected from *field* or real-life settings. Since the original examiner possesses knowledge of the case facts and can observe the demeanor of the suspect during interrogation, this extraneous information might influence his or her scoring or interpretation of the test. Therefore, field studies have used a design in which different examiners, ignorant of the case facts, "blindly" rescore polygraph charts produced by suspects later determined to have been either truthful or deceptive.

Determining Ground Truth

Although free of the disadvantages of the analog design, real-life studies must confront the problem of determining ground truth. Most commonly, confessions have been used as the criterion, either establishing that the person tested was lying, because he or she subsequently confessed, or that some alternative suspect in the same crime, cleared by another's confession, was telling the truth during his or her polygraph test. Unfortunately, relying on confessions to establish ground truth has serious drawbacks as we have seen. The problem is not with the confession per se, but with the consequences of the method used to get confessions.

In CQT studies that have used the confession method to try to estimate polygraph accuracy, because they are obtained pursuant to interrogation after "failing" the CQT, these confessions invariably verify *that* test as accurate. In cases with more than one suspect, such a confession may also clear other suspects; if another suspect has "passed" a CQT prior to the confession, that prior test will also be verified as accurate. All the polygraph errors in which an innocent person failed a test are omitted from such studies because, absent a confession, none of these cases would qualify for inclusion. Similarly, because there would be no confession, all the cases in which a guilty subject erroneously *passed* a test would also be excluded. Thus, confession studies rely on a biased set of cases by systematically eliminating those containing errors and including only those where the original examiner was shown to be correct.

As noted above, polygraph scoring is reasonably consistent from one polygrapher to another. Consequently, when this biased set of confession-verified cases, all chosen in such a manner as to guarantee that the original examiner was correct, is rescored blindly by another examiner, it should be no surprise that the second examiner is also correct. Nothing can be concluded about the accuracy of the CQT from a study like this. The consequence of reliance on the confession criterion is that such studies *must* overestimate the validity of the CQT in the case of both truthful and deceptive suspects.

To determine if they could overcome this problem, Patrick and Iacono carried out a field study with the Royal Canadian Mounted Police in Vancouver, British Columbia.[15] These investigators began with 402 polygraph cases representing all of the cases from a designated metropolitan area during a five-year period. Rather than rely on confessions that were dependent on failing a polygraph test to determine ground truth, they searched police investigative files for ground truth information uncovered after the polygraph test was given, such as non-polygraph-related confessions or statements indicating no crime was committed (e.g., something reported stolen was really lost and subsequently recovered by the owner). These authors found only 1 case out of over 400 that independently established the guilt of someone who took a polygraph test.

The reason why it was not possible to establish independent evidence of ground truth for guilty persons in this study lies in how law enforcement agencies use polygraph tests. Typically, a lie detector test is not introduced into a case until all the leads have been exhausted and the investigation is near a dead-end. If one of the suspects fails the test, it is hoped that the subsequent interrogation will lead to a confession, thereby resolving the issue at hand. However, if the test is failed and there is no confession, there will be no new leads to follow, so the police, assuming that the person who failed the polygraph is guilty, do not investigate the case further. Hence, there is almost no opportunity for additional evidence establishing ground truth to emerge. This disappointing finding indicates that, as polygraphy is now employed in law enforcement, it is virtually impossible to establish ground truth in a manner that is independent of polygraph test outcome. Consequently, the accuracy of the CQT for guilty subjects cannot be reliably determined from the research thus far reported.[16]

If all the suspects in a case pass a polygraph, however, because the guilty person could potentially still be identified through further police work, the file is kept active and additional leads are investigated if they arise. Hence, Patrick and Iacono were able to identify 25 cases where no suspect failed a test but where, for example, subsequent investigation led to confessions from suspects who never took a polygraph test but whose confession established as innocent those who had. Because these confessions were not dependent on someone having failed a CQT, they could be used to establish the accuracy of the CQT for innocent people by having the physiological charts blindly rescored. The results of this rescoring were that 11, or 45%, of these 25 innocent suspects were erroneously classified as deceptive by the CQT.

What would constitute a scientifically credible study of lie detector validity? Suppose that the FBI, with its extensive facilities and its staff of 40-odd polygraphers stationed about the United States, were to undertake to administer polygraph tests to every suspect who could be persuaded to submit to the procedure. As will be seen, the candidates could be assured that the outcome of the test would be kept secret until the case was closed so that negative results would not and could not be used against them. The charts resulting from these tests would not be scored but instead filed away against the day when the true facts of that case became known by means unrelated to the test itself. At that time, when and if the bureau decides that the case can be closed, a panel of criminal prosecutors and defense attorneys would independently evaluate the evidence, and for each suspect tested during the investigation, they would render a judgment as to that suspect's guilt or innocence, using a five-point scale ranging from "certain guilt" to "certain innocence." For suspects who receive a unanimous rating of either certain guilt or certain innocence, the polygraph charts would be blindly scored by polygraphers unfamiliar with the case facts, and their scorings compared against the criterion judgment. When at least, say, 100 confirmed cases of both guilt and innocence had accumulated, the true-positive hit rates (the proportion of guiltys scored as deceptive) and the true-negative hit rates (innocents scored as truthful) would provide a reasonable estimate of lie test validity at least in the FBI context. Unhappily, because polygraphers are themselves persuaded that their tests are nearly infallible, no major police agency in the United States is likely to undertake such an investigation.

Summary

Laboratory studies, because they do a poor job of simulating the high-stakes scenario that exists in real-life criminal investigations, cannot be relied on to estimate polygraph (lie detector) accuracy. Field studies that employ confessions following a failed polygraph also cannot be relied on for this purpose. Credible field studies would be possible, but difficult to implement. To date, no scientifically credible field study of the validity of the CQT in detecting guilty suspects has been accomplished. However, as Patrick and Iacono showed, because police practices differ between cases that yield at least one failed test and those that yield only passed tests, it is possible at least to collect data that bear on the accuracy of the CQT for innocent subjects.

Part II

LIE DETECTION: THE METHODS

> *What we, the American people, are witnessing is the beginning of the end of mankind's search for an honest witness. For the first time in the history of civilization, mankind has the opportunity to prove beyond any reasonable doubt the veracity of his testimony through a generally accepted and scientific [sic] valid examination of his own psyche. God gave us the polygraph.*
>
> —MICHAEL B. LYNCH, in *Polygraph*, The Journal of the American Polygraph Association, 1975

These seven chapters are devoted to a critical evaluation of the various methods that are commonly used for purposes of lie detection. Many polygraphers believe that the examiner is the real "lie detector" and that polygraphers can be trained to be expert practical psychologists able to spot the "symptoms" of deception more skillfully and accurately than the rest of us can. Polygraphers of this persuasion use the polygraph recordings in reaching their diagnosis, but only as an adjunct. They think of themselves rather like experienced medical diagnosticians who rely heavily on the look and feel and history of the actual patient in reaching their conclusions. Chapter 6 considers the logic of this approach and the evidence concerning how well it seems to work.

Most of the more recently trained polygraphers, in contrast, try to focus on the polygraph recordings exclusively. They do not pretend to be especially perceptive in evaluating people nor do they claim that the polygraph itself can detect lying. They contend instead that the polygraph charts provide evidence from which an expert can reliably infer whether the subject was being truthful or deceptive. However, the polygraph recordings will be informative only if the subject has been asked the right questions in the right sequence and after he has been psychologically prepared in the right way. Polygraphers use several different question formats in lie detection; which one is used with a particular subject will depend on the examiner's training and the problem at hand. The principal variants are each discussed in this section.

Whichever question format is employed, we are dealing with a *polygraphic lie test* only if the results of the examination are determined entirely by the polygraph recordings, so that a second polygrapher, seeing only the charts, would be likely to arrive at the same conclusion. If other information is allowed to influence the results, if the examiner is the "lie detector" (Chapter 6), then we are dealing with a *clinical lie test*, no matter which of the question formats has been used. When the examination is evaluated subjectively in this manner, we cannot tell what role the actual polygraph results played in the outcome; we cannot then speak of the validity of the polygraph test but rather only of the validity of that particular examiner.

One of the paradoxes of polygraphic lie detection—the fact that the examiner must succeed in deceiving the subject in order not to be deceived *by* the subject—is reviewed at the end of Chapter 10, together with a table summarizing what is known about the accuracy of the several lie detection methods.

The minor vogue in some lie detection circles is for a relatively new technique that is said to be able to assess "stress"—and from these assessments, to infer deception—in the *voice* of the subject. These "voice stress analyzers" can be used in place of—or together with—the conventional polygraph in any of the various interrogation modes. The curious and, I think, scandalous story of the voice stress analyzers is told in Chapter 11.

Because no definitive assessment of lie detector accuracy has been accomplished, for the reasons that are detailed in Chapter 5, courts, government officials, and citizens in general need help in deciding what to make of (and what to do about) this intimidating technology. I try in this book to provide facts and explanations sufficient to permit the reader to form an independent opinion; but are my facts and explanations to be trusted? It is natural, in such important matters, to want a second opinion. My colleague, Bill Iacono, and I have recently elicited the opinions about polygraphy and lie detection of two groups of scientists who are knowledgeable about matters of this kind. The methods and results of these two surveys are presented in Chapter 12.

Chapter 6

THE CLINICAL LIE TEST:
THE EXAMINER
AS "LIE DETECTOR"

When the eyes say one thing, and the tongue another, a practical man relies on the language of the first.

—EMERSON

I do not distinguish by the eye but by the mind, which is the proper judge of man.

—SENECA

"Sure, a lie's a lie," said Mr. Hennessy. "I always know whin I'm lyin'." "So do I," said Mr. Dooley.

—F. P. DUNNE

Mary St. Claire was killed during a drunken party in the woods, a party shared with Mary's stepdaughter, Charlene, and with John Fontaine. Death was caused by the blow of an ax that nearly severed the head from the body. John was arrested the next day and for weeks he stood mute, refusing to explain what had happened. At last he instructed his attorney to enter a plea of guilty to second-degree murder, although the account he then gave of the incident specified that Charlene had struck the fatal blow. "Charlene has a family and a job and I don't, so it is better if I take the blame." His attorney was reluctant to plead his client guilty if he was truly innocent and so he came to me for help. Would I give John a lie detector test to establish who, in fact, had chopped off Mary's head?

I was careful to explain that not even the elaborate apparatus in my laboratory could actually detect lying. But I knew from the published research that untrained persons can distinguish liars from the truthful with about 70% accuracy. Since I am a clinical psychologist with considerable experience, I assumed that I could do at least that well. Moreover, I could use all the tricks of the polygraph examiner: the special interviewing tactics, the scientific stage props. If John was guilty, I might be able to get him to confess and remove the ambiguity. If he did not confess, in spite of all my efforts, then perhaps we could be more confident that his story was the truth.

The night John was brought to my laboratory by the sheriff's deputies, thousands of dollars' worth of electronic equipment were whirring and humming. I unerringly determined, seemingly from the polygraph, which card it was that John had selected from my (fixed) deck. I spent more than an hour in the pretest interview, preparing the ground. After the first charts had been run, I tightened the psychological thumbscrews:

> John, I'm getting reactions here that show you aren't telling the whole truth about what happened that night. Can you think of any reason why you might be giving these reactions?

John looked bewildered and made no effort to modify his story or to explain these alleged "reactions." We ran another chart.

> John, I realize that you were drunk, that you had every reason to get mad at Mary, the way she was behaving. And I know how hard it is, when you're sober, to look back at the things you did when you were drunk. But the polygraph says that you haven't told me the whole truth about this. If you want your lawyer to let you plead guilty, at least do him the favor of letting him know what really happened.

John stuck with his previous account; Mary and Charlene were fighting and he tried to break it up. Mary fell and Charlene seized the firewood ax and struck her where she lay.

I told the lawyer what I had done, that I had not managed to shake John's story, that my clinical opinion was that he was telling me the truth. Two weeks later, back in jail, John told his attorney that he had swung the ax himself. I still do not know ground truth with any certainty. Had John's stoicism fooled me or was this new confession merely John's way of implementing his decision to take the blame? We are all of us "human lie detectors"; I had thought I was more skillful than most. I could still insist that my original opinion was correct, but more than 40 years' experience as a psychologist has taught me a little humility. People are complicated; it is dangerous to be too sure of what you think you know about them.

Subjective Scoring and "Behavior Symptoms"

What I did with John Fontaine is an example of a clinical polygraph examination. I, the examiner, was the lie detector. Traditional polygraphers, like John Reid or Richard Arther, spoke of the results of such an interrogation as a "diagnosis." This is an appropriate term, because the examiner is acting rather like a physician does in examining a patient, taking all sources of information into account, including clinical impressions based on his experience; he combines them according to some subjective formula and arrives at his best guess as to what is going on. I thought hard about John's story of what happened on the fatal night, about whether his account seemed consistent and plausible. I studied his behavior for more than two hours, my clinical "radar" turned fully on, looking for signs that he was covering up or inventing as he went along. I applied pressure to see if he would modify his story in order to make it more convincing or to explain away the indications of deception that I pretended to see in the polygraph. I did not put much weight on the actual polygraph tracings because I did not think that they contained much useful information. A clinical polygrapher would pay more attention to the charts than I did, but even he cannot know, either in general or in a particular case, how strongly he is being influenced by the actual polygraph data, because he arrives at his diagnosis subjectively; he exposes himself to all the available information and the diagnosis simply emerges from the mysterious computer of his mind. Clinical judgments, by definition, do not employ specific rules for combining the available information. The process is partly deductive and partly oracular.

It is therefore difficult to discuss the theory underlying the subjective scoring methods of Keeler or Reid or Arther. If the principles according to which the different types of evidence are supposed to be combined had ever been clearly specified, then an "expert" would not be required. The results of an examination could be determined by a clerk following a rule book, just as a Backster-trained polygrapher can score the charts from a polygraphic lie test administered by a colleague. Perhaps we can infer the essentials of the theory from a closer study of the way a Reid-type examination is conducted. The examiner does everything he can to impress the subject with the impartiality, professionalism, and the scientific basis of the procedure. The waiting room may be hung with the examiner's diplomas and furnished with reading material attesting to the respectability and validity of polygraph methods. The examiner will behave toward the subject in a polite and professional way. The examining room is carpeted and quiet and dominated by the impressive aspect of the polygraph itself.

The subject will not be connected to this apparatus until a lengthy pretest interview has been conducted. During this interview and the polygraph test itself, the examiner will be studying his subject, observing his demeanor, his actions and appearance, his "behavior symptoms." If a secretary is available in the waiting room, she may be instructed to note how the subject comported himself while awaiting the polygrapher. A well-equipped examining room will contain a one-way mirror, disguised as a picture or (in at least one room I have visited) as a tropical fish tank set into the wall. The examination may then be interrupted at some point and the subject left to stew alone while the examiner observes him from the adjacent room.

In 1953, Reid and Arther reported the results of a five-year tabulation of "behavior symptoms," which they said discriminated between lying and truthful subjects during a polygraph examination.[1] In the last edition of the Reid and Inbau textbook, a section entitled "Symptoms of Lying" explains that the deceptive subject is reluctant to take the test and may postpone or be late for his appointment.[2] Appearing nervous, resentful, aggressive, "appearing to be in a shocked condition," exhibiting "mental blocks," having a dry mouth or a gurgling stomach, refusing eye contact, moving restlessly, appearing "overly friendly or polite," describing himself as religious, complaining of pain from the blood pressure cuff, being eager to finish the examination and leaving promptly—all these are behavior symptoms of the liar. Under "Symptoms of Truthfulness," Reid and Inbau list an eagerness to take the examination, a feeling of confidence in the test, an attitude of sincerity and straightforwardness, an appearance of composure, and behaving in a cooperative manner. To emphasize that these are diagnostic symptoms and not merely tendencies or trends, Reid and Inbau include a third list of behaviors said to be "Common to both

Liars and Truthtellers" that lack the diagnostic specificity of the other two tabulations. Other examiners develop their own lists of behaviors that they think are revealing (Arther has been especially creative in this regard), and any examiner will diagnose as deceptive a subject who he thinks is attempting to control his polygraphic reactions by moving, breathing erratically or too regularly, or twitching the muscles in the arm carrying the blood pressure cuff.

Most polygraphers of the Keeler–Reid–Arther persuasion consider the *posttest interrogation* to be an essential component of the examination. If the examiner's impressions up to that point have led him to believe that the subject is being truthful, and if this view seems to be confirmed by a lack of strong physiological reaction to the relevant questions, then the interrogation may be dispensed with. But when the polygrapher thinks that the subject is lying or may possibly be lying, then he will remove the polygraph attachments, seat himself facing the subject, knee to knee, and begin: "The polygraph shows that you are not telling the truth." Or, as in the Peter Reilly case, "Pete, I think you got a problem, I really do.... These charts say you hurt your mother last night." (We shall have more to say about this notorious case in Chapter 17.)

The interrogation obviously is calculated to elicit a host of additional "behavior symptoms." It is also calculated to elicit a confession. Most old-school examiners of the type we are concerned with here, "human lie detectors," are primarily interrogators and their real goal, the prize that most clearly demonstrates their skill and the potency of their technique, is a confession. And the polygraph examination from first to last is a powerful inducement to confession.

If the subject does not confess, however, then the examiner must render his decision. He considers the polygraph charts, the behavior symptoms he has noted, the case facts as he understands them, and the subject's explanations or alibi as elicited during the pretest interview and the posttest interrogation. There may be other relevant information available as well. If this is just a routine screening of a bank employee, then there will be less prior expectation of deception than if this subject was the only person known to have had access to the vault from which a sum of money has been stolen. If 20 employees had access to that vault and 19 remain to be tested, the examiner may be less inclined to render a "deceptive" diagnosis than if this man appears to be the only suspect. If the case involves one person's word against another's and the other person has already been diagnosed as truthful, then it will be harder for the examiner to classify the present subject as truthful. Finally, of course, there are the irrelevant and even unconscious sources of bias that the examiner will strive to ignore: his own attitude toward the subject, the subject's age, sex, appearance, race, ethnic or cultural background. There are no rules for selecting, weighting, or combining these various pieces of evidence. Reid and Inbau assert in their textbook that one should not place "sole or even major reliance" on behavior symptoms, yet, testifying in a courtroom in Toronto in 1976, John Reid said that the identification of behavior symptoms "is a big part of our course, much more than running charts, they don't mean anything almost. You can put a small boy in to do that. This is unimportant."[3] Most old-school examiners would probably agree that this central question—how to weigh and combine the available information to produce a final diagnosis—is decided according to each individual polygrapher's experience and intuition.

Assumptions of the Clinical Polygraph Examination

Polygraphers have not provided an agreed-upon name for the kind of examination taught by the Keeler, Reid, and Arther schools, in which the examiner himself serves as lie detector and the polygraphic information is combined with impressions of behavior and other data in arriving at a diagnosis. But we need to distinguish this approach from the Backster method, which attempts to let objective scoring of the charts alone determine the result. Both are psychological procedures aimed at arriving at a judgment about the subject's psychological state. The Backster technique is arguably a psychological test, but the Keeler–Reid–Arther procedure is clearly not a test at all, although it is commonly referred to in this way. A psychiatrist's interview, a physician's physical examination, a jury's evaluation of a witness, a subjectively scored polygraph examination, these all can be described as clinical assessments rather than as tests. Although the theory of the *clinical polygraph examination* has never been spelled out in the literature of the field, let us examine the assumptions on which the procedure seems to rest.

ASSUMPTION 1. *The examiner will be able to convince every subject that the results of his polygraph examination are virtually certain to be accurate.*

Unless a truthful subject has real faith in the "test," he is likely to show some of the behavior symptoms attributed to the liar and also to show strong polygraphic responses to the relevant questions. The procedure clearly assumes that the deceptive subject will feel stress and that the truthful subject will not; this is plausible only if all subjects genuinely believe that the "test" will reveal the truth.

Can polygraphers persuade all their subjects that the polygraph method is nearly infallible? In the United States, the examiner has 70 years of mythology working for him. I was once consulted by a young man who had failed a lie test relating to a theft: His faith was so strong that he believed the polygraph results rather than his own memory. He hurried home from the examination to search his apartment for the missing $400, thinking that he must have taken it during a "blackout" since he had no recollection of it. He came to me fearing that he was losing his mind. I have talked with another man who knew he had not stolen the money in question but who concluded that the polygraph must have detected some deeper stain on his character, that he would not have failed the test unless he was in fact "bad" in some way. The transcripts quoted in Barthel's *A Death in Canaan* dramatically illustrate how 18-year-old Peter Reilly's faith in the machine (and in fatherly Sergeant Kelly, the polygrapher) led him to believe against the evidence of his senses that he had murdered and mutilated his own mother.

To augment this kind of faith, standard polygraph procedure uses a stimulation procedure, or "stim test," in which the examiner pretends to be able to determine from the polygraph charts which card the respondent has chosen from a deck. Reid and Inbau explain:

> The cards are arranged and shown to the subject in such a way that the examiner will immediately know which card has been picked by the subject. The reasons for this are (1) the card test record itself may not actually disclose the card "lie"; (2) the primary purpose of the card test is the "stimulation" effect that results ... by reason of the subject's belief that his card test "lie" was detected, and obviously, unless his chosen card is correctly identified after the card test, the stimulation effect is lost completely.[4]

Oftentimes the subject's largest polygraphic response will follow the question referring to the card he actually chose, so that the examiner's guess, based on the polygraph, would be correct. A number of studies confirm this.[5] In a study using charts obtained during polygraphic interrogation of actual criminal suspects, one group of investigators were able to identify the correct card 55% of the time. Laboratory studies have reported success rates from about 30% to 73%. As Reid and Inbau point out, however, since the object of the "stim test" is to convince the subject that the polygraph can determine when he is or is not lying, being correct only half or two-thirds of the time is not good enough. Therefore, one must resort to a stacked deck.

Many polygraphers now use a different procedure that they believe to be less deceptive. The subject is asked to pick a number, say, from 1 to 7. That number is then written large on a piece of paper and hung on the wall where both the subject and examiner can see it. As in the card test, the subject is told to reply to "No" to each question of the form, "Did you choose number X?" The subject is also told that the purpose of this procedure is to calibrate the polygraph, "so that I can determine what your polygraph responses look like when you are lying and when you are telling the truth." The ability of the procedure to inspire confidence in the subject depends on this latter statement, which, of course, is untrue and misleading. Selecting the number, revealing the choice to the examiner, seeing it prominently displayed when the question is asked, all this makes the chosen digit significant to the subject and leads him to react more strongly to that one than to the others. None of this has anything to do with lie detection and it is simply deceptive to suggest that the subject is showing a pattern of reaction that the examiner now knows to be the way he responds when he is lying. Since there seems to be as much charlatanry in this method as in Keeler's or Reid's, it does not seem to be a real improvement. If one is going to deceive the subject, why not use the deception that is most effective?

Against those who object to these "stim tests" on ethical grounds, it can be argued that the truthful subject will be safer (more likely to "pass" the test) if he can be deceived in this way and that the deceptive subject has no inherent right to be treated truthfully by others. A more serious objection, however, is that it is risky to base a widely used procedure on a kind of trick or fraud that, inevitably, some people will learn about, perhaps especially those people tuned in to the underworld grapevine over which such information is disseminated. Someone who has read this book, for example, will be less likely to be impressed by the card test and also less likely to take the control questions as seriously as the relevant questions, having learned that his or her fate may depend on not giving large responses to the latter. As an eminent Canadian jurist wrote after hearing testimony from both Reid and Arther on the methods they espouse, "Convincing a subject that the machine is infallible which is certainly not true and using a rigged card test for this purpose is a formula no competent scientist would accept."[6]

ASSUMPTION 2. *If all subjects have faith in the procedure, then there are certain behavior symptoms that will be shown only by truthful subjects and other, different symptoms that only deceptive subjects will manifest.*

This is the sort of claim that makes a psychologist's hair stand on end. Does it really require years of postgraduate study to realize that every "symptom" some people show when they lie, other people will sometimes display when they are being truthful? Just a few hours alertly spent at a poker table will reveal that whatever Smith does when he's bluffing, Jones may only do when he is holding a strong hand. Reid and Arther claim to have demonstrated the validity of their symptom lists in a five-year study published in 1953.[7] Frank Horvath, one of Reid's colleagues, published a somewhat less impressionistic study in 1973 that seemed to confirm these claims, at least in part; deceptive subjects more often showed "typical liar" symptoms than did truthful subjects and vice versa.[8] But Horvath's data consisted of notations made by Reid examiners in the course of routine polygraph tests. Horvath acknowledges that these observations were inevitably contaminated in such a way as to tend to produce the expected findings. To take just one example, the "liar" symptom of "poor eye contact" was noted for half of the deceptive subjects but for none of the truthful ones. Classifying eye contact as "good" or "bad" involves some subjectivity. An examiner trained to regard poor eye contact as symptomatic of lying will be unlikely to classify a subject's eye contact as "poor" if he believes on other grounds that the subject is being truthful. And the examiner has many other grounds for forming an opinion during the interview: the strength of the evidence against the subject, the subject's own story, his demeanor and other behavior. To obtain truly objective and independent estimates of how consistently "lying symptoms" indicate a deceptive subject, or what proportion of truth-tellers have good eye contact, a cooperative attitude, and are "genuinely" but not "overly" friendly, one would want to videotape interviews and examinations and have different judges rate the various symptoms under circumstances designed to minimize the tendency for judges to see what they think they are supposed to be seeing. Since one would require thousands of taped interviews in order to yield a sufficient number of cases confirmed by subsequent evidence, it is unlikely that any trained scientist would undertake such an investigation merely to confirm that Reid and Arther's claims are simplistic and greatly overstated. Thus the cited studies stand alone in the field and the only antidote available is common sense—which, it appears, is too weak a potion.

Law professor E. A. Jones, an experienced labor arbitrator, offers a more sophisticated view of the dependability of "behavior symptoms":

> Anyone driven by the necessity of adjudging credibility who has listened over a number of years to sworn testimony, knows that as much truth has been uttered by shifty-eyed, perspiring, lip-licking, nail-biting, guilty-looking, ill-at-ease fidgety witnesses as have lies issued from calm, collected, imperturbable, urbane, straight-in-the-eye perjurers.[9]

Similarly, after hearing extended testimony from both Reid and Arther, the leading exponents of the clinical lie test, consider the reaction of Mr. Justice Morand:

> I was amazed at the naive and dogmatic pronouncements by polygraphers concerning interpretations of behavior, many of which were founded on the assumption that a reluctant subject or an opponent of the polygraph is probably a liar. Little or no account was taken of the variations in psychological reactions which require great flexibility in assessment of individuals.[10]

I have not disproved Assumption 2; that would be an expensive and thankless undertaking. But I have tried to show that this assumption is so "naive and dogmatic" that a considerable burden of proof lies on those who would have us believe that certain behavior symptoms almost always reveal whether someone is truthful or lying. Neither the Reid and Arther study nor the later report by Horvath was properly designed so as to provide the proof required.

This is not to say that observers cannot sometimes separate the truthful sheep from the deceptive goats just by observing them during an interrogation. In a laboratory study using a mock crime situation, Kubis found that his experienced examiners could correctly classify 65% of their subjects, one-third of whom were "guilty," one-third "accomplices," and one-third "innocent," just by observing their interrogations. On a small sample of criminal suspects, Barland made diagnoses based on interview behavior that were correct 69% of the time.[11] It might seem safe to assume that a perceptive observer, an experienced police detective, for example, could interrogate 100 criminal suspects, half of whom are truthful, and then classify them as truthful or deceptive with about 70% accuracy. More recently, however, psychologists Paul Ekman and Maureen O'Sullivan found that only U.S. Secret Service agents attained this accuracy in systematic testing. "All of the other professional groups concerned with lying—judges, trial attorneys, police, polygraphers who work for the CIA, FBI, or NSA (National Security Agency), the military services, and psychiatrists who do forensic work—did no better than chance."[12] There is no evidence that a study of Reid and Inbau's lists of behavior symptoms would improve on this level of accuracy. My guess is that the opposite would happen, that what now one sees "as through a glass, darkly," one might then not see at all.

ASSUMPTION 3. *When a behavior symptom cannot be directly observed but must be inferred by the examiner (e.g., "subject is nervous" or "subject is sincere"), all examiners will be skillful enough to make correct inferences in dealing with all varieties of subjects.*

Assumption 3 requires that examiners will be able to make correct inferences about each subject's attitudes and feelings—gauge nervousness, identify sincerity, measure the subject's confidence—after a six-week course and a six-month apprenticeship. These are judgments that we all make frequently in ordinary life, but evidence shows that we are not nearly as accurate as we think. Studies of stage fright, for example, show that ratings by the audience of a speaker's nervousness or confidence have only the roughest sort of relationship to how the speaker really feels, as measured either by his subsequent report or by polygraph recordings made while he is speaking. Experienced psychotherapists may spend dozens of hours in intimate conversation with a patient before they can read with confidence those feelings that the patient tries to hide. If their perception sharpens after all those hours, it is because they learn the idiosyncratic meaning of the "behavior symptoms" shown by that individual patient. Assumption 3 also has to be rejected, not so much because the typical polygrapher has negligible psychological training, but because not even a psychologist could make these subtle judgments with high accuracy on the brief acquaintance provided by a polygraph examination.

ASSUMPTION 4. *Given that the subject believes that the polygraph "works," then all deceptive subjects will be more aroused by the relevant questions than by irrelevant or control questions and all truthful subjects will be equally or less aroused by the relevant or "Did you do it?" questions than by the other questions asked of them.*

Assumption 4 concerns the polygraph portion of the examination, how strongly truthful and deceptive subjects can be expected to react, physiologically, to the various types of questions asked. The plausibility of this assumption will depend on which type of question—which test format—is to be used, and we shall consider this assumption in relation to these various formats later on. But the first three assumptions of the clinical polygraph examination are each so implausible that—at the very least—one should require strong experimental evidence before accepting claims that diagnoses based on these assumptions are 90% to 99% accurate.

Validity of the Clinical Lie Test

In view of the millions of clinical lie tests that have been administered to date, it is surprising that only one serious investigation of the validity of this method has been published, Bersh's 1969 Army study.[13] Bersh wanted to assess the average accuracy of typical Army polygraphers who routinely administered clinically evaluated lie "tests" to military personnel suspected of criminal acts. He obtained a representative sample of 323 such cases on which the original examiner had rendered a global diagnosis of truthful or deceptive. The completed case files were then given to a panel of experienced Army attorneys who were asked to study them unhindered by technical rules of evidence and to decide which of the suspects they believed had been guilty and which innocent. The four judges discarded 80 cases in which they felt there was insufficient evidence to permit a confident decision. On the remaining 243 cases, the panel reached unanimous agreement on 157, split three-to-one on another 59, and were deadlocked two-to-two on 27 cases. Using the panel's judgment as his criterion of ground truth, Bersh then compared the prior judgments of the polygraphers against this criterion. When the panel was unanimous, the polygraphers' diagnosis agreed with the panel's verdict on 92% of the cases. When the panel was split three-to-one, the agreement fell to 75%. On the 107 cases where the panel had divided two-to-two or had withheld judgment, no criterion was of course available.

Bersh himself pointed out that we cannot tell what role if any the actual polygraph results played in producing this level of agreement. In another part of that same Defense Department study, polygraphers like those Bersh investigated were required to "blindly" rescore one another's polygraph charts in order to estimate polygraph reliability. The agreement was better than chance but very low. As these Army examiners then operated (they have since converted to the Backster method, which is more reliable), chart scoring was conducted so unreliably that we can be sure that Bersh's examiners *could not* have obtained much of their accuracy from the polygraphs: validity is limited by unreliability. But, although these findings are a poor advertisement for the polygraph itself, can they at least indicate the average accuracy of a trained examiner in judging the credibility of a respondent in the relatively standardized setting of a polygraph examination?

Bersh's examiners based their diagnoses in part on clinical impressions or behavior symptoms, which, we know from the evidence mentioned above, should not have permitted an accuracy much better than chance. But they also had available to them at the time of testing whatever information was then present in that suspect's case file: the evidence then known against him, his own alibi, his past disciplinary record, and so on. In other words, the polygraphers based their diagnoses in part on some portion of the same case facts that the four panel judges used in reaching their criterion decision. This contamination is the chief difficulty with the Bersh study. When his judges were in unanimous agreement, it was presumably because the evidence was especially persuasive, an "open-and-shut case." It may be that much of that same convincing evidence was also available to the polygraphers, helping them to attain that 92% agreement. When the evidence was less clear-cut and the panel disagreed three-to-one among themselves, the evidence may also have been similarly less persuasive when the lie tests were administered—and so the polygrapher's agreement with the panel majority dropped to 75% (note that the average panel member also agreed with the majority 75% of the time). An extreme example of this contamination involves the fact that an unspecified number of the guilty suspects confessed at the time of the examination. Because the exams were clinically evaluated, we can be sure that every test that led to a confession was scored as deceptive. Since confessions were reported to the panel, we can be sure also that the criterion judgment was always guilty in these same cases. Thus, every lie test that produced a confession was *inevitably* counted as an accurate test, although, of course, such cases do not predict at all whether the polygrapher would have been correct absent the confession. That the polygraph test frequently produces a confession is its most valuable characteristic to the criminal investigator, but the occurrence of a confession tells us nothing about the accuracy of the test itself.

Thus, the one available study of the accuracy of the clinical lie test is fatally compromised. Because of the contamination discussed above, the agreement achieved when the criterion panel was unanimous is clearly an overestimate of how accurate such examiners could be in the typical run of cases. When the panel split three-to-one, then at least we know that there was no confession during the lie test or some other conclusive evidence available to both the panel and the examiner. The agreement achieved on this subgroup was 75%, equal to the panel judges' agreement among themselves. As we have seen, Bersh's examiners could not have improved much on their clinical and evidentiary judgments by referring to their unreliable polygraphs.

Verdict

Whatever the question format may be, whether the test is for general screening purposes or deals with a particular crime or other specific issue, many lie tests still are scored according to the examiner's subjective appraisal of the subject's truthfulness. In criminal cases, this appraisal will be influenced by the examiner's knowledge of the case facts; such knowledge will usually depend in part on rumor or hearsay, it will often be incomplete, and it will sometimes be inaccurate. In all cases, the examiner's appraisal will be influenced by his clinical impressions of the subject and whatever behavior symptoms he thinks he has observed. Most of us think we can assess credibility with some accuracy in this way, but there is no reason to suppose that polygraphers, as a group, are especially skillful. In all cases, the examiner's appraisal may be influenced by extraneous factors: the subject's race, age, sex, or social class; whether the subject or someone else is paying for the test; the examiner's personal stake, if any, in the outcome.

Referring to clinical lie tests of this type, Justice Morand concluded that they involve

> only an assessment of credibility based on observations that jurists have been making for years, by persons who are untrained in conducting psychological examinations and who apparently accept naive and simplistic criteria of what is deceptive behavior and what is not.[14]

We may agree that some people are more gullible than others, but we should not concede that anyone, any group of self-styled professional *Menschenkenner* or polygraphers, is so especially discerning and perspicacious that we would be willing to surrender the important responsibility of credibility assessment primarily to them. Unless the result of a polygraph examination is based solely on the charts—the polygraph recordings—then the result will be influenced to an unknowable extent by the examiner's clinical impressions, evidentiary judgments, personal attitudes, and expectations. The clinical judgment of a polygraph examiner is no more valid than that of any other observer, just as subject to bias and prejudice, and probably wrong nearly 50% of the time.

Chapter 7

THE RELEVANT/IRRELEVANT (R/I) TEST

When we lie, our blood pressure goes up, our heart beats faster, we breathe more quickly (and our breathing slows once the lie has been told), and changes take place in our skin moisture. A polygraph charts these reactions with pens on a moving strip of graph paper.... The result is jagged lines that don't convey a lot to you. But ... an examiner can tell from those mechanical scribbles whether or not you've spoken the truth.

—CHRIS GUGAS, polygrapher, *The Silent Witness*, 1979

The widespread use of polygraphic interrogation methods is not based on public acceptance of the idea that polygraphers are better human lie detectors than are judges or juries or personnel interviewers or, indeed, than people in general. Suppose that, in a criminal trial, the prosecution called to the stand Mr. Reid or Dr. Larson or Father Summers and sought to have any one of them accredited as an expert witness without benefit of the polygraph:

> Your honor, Mr. Reid has many years of experience in interrogating criminal suspects. He has spent two hours interviewing this defendant. He has reviewed the case facts, heard the defendant's alibi, asked him various searching questions, and closely observed his demeanor and behavior. As an expert in the diagnosis of deception, Mr. Reid is prepared to testify that, in his expert opinion, this defendant is lying when he denies his guilt in the present matter.

One does not have to be learned in the law to predict how most judges would rule on this motion. Yet, there is currently a trend for trial judges in some U.S. jurisdictions to qualify polygraph examiners as expert witnesses under special circumstances. Clearly the difference has to do with the polygraph itself, this mysterious machine that draws those mystic markings on the chart paper, and with the aura of scientism and expertise that surrounds the person who can operate the machine and read the markings.

Therefore, it is necessary to consider separately the validity that can be achieved in detecting deception strictly on the basis of the polygraph charts. In the next three chapters we shall examine the *polygraphic lie tests* in common use, the assumptions on which they are based, and the evidence, if any, of their validity. The only way to assess the validity of a polygraphic lie test, independent of clinical impressions, is to have the polygraph charts scored "blindly" by a second polygrapher who did not observe the subject and is unfamiliar with the case facts. Since polygraph charts are seldom scored this way in practice, however, it should be remembered that most lie detector "tests" are really clinical examinations, subject to the problems and limitations discussed in the previous chapter.

In addition to this requirement of blind scoring, validity studies must be conducted in the field, in real-life testing situations. Volunteer subjects in laboratory experiments are not under the same emotional pressure that affects criminal suspects or persons being screened for employment purposes. Deceptive subjects may be less reactive in the laboratory than when lying about real crimes at the police station. Truthful subjects will almost certainly be more reactive in the field situation and, for this reason, more likely to be misclassified by the polygraphic lie test.

Assumptions of the R/I Test

In the parlance of polygraphy, a relevant question is the "Did you do it?" question: Did you kill Fisbee? Did you fire the bullet that killed Fisbee? and so on. An irrelevant question, sometimes called a "norm," will be both unrelated to the matter under investigation and also nonstressful: Is today Tuesday? Are you in Chicago now? All polygraphic lie tests also employ "control questions," but these are less easily defined. Confusingly, the relevant/irrelevant, or R/I, test, Larson's original invention and the forerunner of them all, uses "control" questions that are entirely different in function than those used in the various "control question tests" to be discussed later. In the R/I test, the control question is irrelevant but provocative: "Have you been drunk at any time during the past year?" or "Do you remember my name?" The sole purpose of these questions is to produce some sort of polygraphic response and thus prove that the subject is capable of responding under the present conditions. If the subject has responded to some of the previous questions on that same chart, the control question may be omitted.

A typical relevant/irrelevant test will consist of three relevant questions, each preceded and followed by irrelevant questions, with an evocative control question at the end of the list. The American Polygraph Association advocated the name "general series test" for this venerable technique, a label that has the disadvantage of being devoid of meaning. "Relevant/irrelevant" is a more straightforward name that served well enough for 70 years and will serve well enough here. If the subject shows a strong polygraphic reaction to some or all of the relevant questions, but not to the irrelevant questions, then his answers to the relevant questions are classified as deceptive. Most psychologists will find it hard to credit that so simple minded a procedure has been in constant, serious use since the 1920s. But since the R/I method is still being taught and is widely used, we must consider it seriously. The assumptions on which the R/I test is based are these:

ASSUMPTION 1. *A guilty subject whose relevant answers are lies will be more aroused by the relevant than by the irrelevant questions and this difference will be revealed by his responses on the polygraph.*

The first assumption is certainly plausible but it is important to see that it cannot be taken as certain in all cases. Some individuals may be so fearless that they are not worried about the consequences of the test or so lacking in conscience that they are not aroused by references to their crime or by the act of lying. A guilty subject may become habituated to such references as a result of prior questioning or may possess unusual emotional control with which to steel himself against reacting. One would expect most guilty suspects to show greater arousal to the relevant questions, but no psychophysiologist would expect this to happen without fail.

ASSUMPTION 2. *An innocent subject who is answering truthfully will not be disturbed by the relevant questions and will show no more reaction to them than to the irrelevant questions.*

The second assumption is wildly implausible. One has only to imagine oneself on trial, testifying in one's own defense, and the moment arrives when counsel asks the critical, "relevant" question: "Did you commit the grave crime of which you now stand indicted?" and one must face the jury, eye to eye, and answer "No!" How would you be feeling at that moment? What would the polygraph pens be saying about your heartbeat, your breathing, and the sweating of your palms? But suppose you *are* innocent, telling the truth? Would that fact armor you with confidence so that you could reply to "Are you guilty?" as calmly as you had previously stated your name and address? How could two generations of polygraphers have been unable to see that the relevant question is a stimulus as threatening and arousing for the innocent as for the guilty, that the truthful denial can be as emotional as a false one?

Adherents of the R/I technique will reject this witness-box analogy as inappropriate and misleading. In the privacy of the polygraph room, using "stim tests" and other methods, the examiner will be able to inspire the subject with such respect for the power and accuracy of the procedure that the truthful subject, having nothing to fear, will be unmoved by the relevant questions. We have seen this assumption before in the discussion of the clinical polygraph test in the previous chapter. Most examiners who use the R/I format also use clinical assessment. But polygraphers *cannot* convince all subjects that the polygraph is infallible. Moreover, "Are you guilty?" will continue to be a significant, arousing stimulus for some people even when they feel certain that their innocent denial is about to be confirmed by the magic machine.

Validity of the R/I Test

So much for theory and common sense; what is the evidence? It is astonishing to discover that, in 70 years of use prior to 1997, the *only* published studies assessing R/I test accuracy using "blind" evaluations of charts obtained from criminal suspects were one described briefly by Larson[1] in 1938 and another, by Horvath,[2] in 1968. Larson asked nine judges to read the charts obtained from 62 suspects. Only 1 of the 62 suspects had actually lied and yet the number scored as deceptive by the nine judges ranged from 5 to 30. This amount of disagreement among the nine judges indicates poor reliability. The average judge scored about one-third of the innocent suspects as deceptive, which means that two-thirds of these innocents failed to give large reactions to the relevant questions and were scored as truthful, just as Assumption 2 demands. One might have thought that Assumption 2 would nearly always be wrong and that most subjects would fail the R/I test whether innocent or guilty. That is in fact what Horvath reported; *all* of his innocent suspects were erroneously classified as deceptive by the R/I test, whereas Larson's earlier study reported only 33% false-positives. We should not put too much faith in the exact percentage of errors found but we can say that, just as common sense would predict, a high proportion of innocent subjects *will* "fail" the R/I test. Quite recently the Raskin group of lie detector advocates published the results of a mock crime laboratory study[3] in which the R/I method classified all 15 of the "guilty" suspects as deceptive but at the expense of identifying only 3 of the 15 "innocent" subjects as truthful.

Verdict

The relevant/irrelevant form of lie test is based on a flagrantly implausible assumption. Common sense indicates that the R/I test will be strongly biased against the truthful subject, and the three bits of evidence available, one from the inventor of the R/I procedure himself, support this assessment.

Chapter 8

THE CONTROL QUESTION
TEST (CQT)

Can you nominate in order now the degrees of a lie?

—Shakespeare, *As You Like It*

It is manifest that man is ... subject to much variability.

—Darwin, *The Descent of Man*

The polygraphic lie test most commonly used in criminal investigation is the method developed by John Reid and known in the trade as the Control Question Test. Backster's Zone of Comparison method is a popular variant of the CQT that differs slightly from Reid's in the arrangement of questions and the methods of scoring. The important difference between the modern CQT and Reid's method is that the CQT is scored relatively objectively and the diagnosis is based on that numerical scoring, eschewing behavior symptoms and clinical impressions.

The format of the CQT is illustrated in Table 8.1. The first question is the familiar irrelevant type. Question 2 is relevant in substance but is not used in scoring the charts; it is called a "sacrifice relevant." Question 3 is an "outside issue" question designed for the situation in which the subject might be afraid that the interrogation will stray into an area about which he has real concern. If he seems disturbed by this question, testing will be postponed until he can be convinced that the only questions asked will

Table 8.1. Example of a Control Question Test Administered
by D. Raskin to a Defendant Accused of Homicide by Stabbing[1]

1.	Were you born in Hong Kong?	Yes	(Irrelevant)
2.	Regarding the stabbing of Ken Chiu, do you intend to answer truthfully every question about that?	Yes	(Sacrifice relevant)
3.	Do you understand that I will ask only questions we have discussed?	Yes	(Outside issue)
4.	During the first 18 years of your life, did you ever hurt someone?	No	(Lie control)
5.	Did you cut anyone with a knife on Dumfries Street on January 23, 1976?	No	(Relevant)
6.	Before 1974 did you ever try to seriously hurt someone?	No	(Lie control)
7.	Did you stab Ken Chiu on January 23, 1976?	No	(Relevant)
8.	Is your first name William?	Yes	(Irrelevant)
9.	Before age 19, did you ever lie to get out of trouble?	No	(Lie control)
10.	Did you actually see Ken Chiu get stabbed?	No	(Relevant)

be those previously reviewed with him. The three relevant questions, numbers 5, 7, and 10, all refer to a certain fatal stabbing for which this subject has been indicted. Question 8 is another irrelevant, interpolated mainly to provide a rest stop between the more arousing questions that precede and follow it.

Let us now consider Questions 4, 6, and 9, the "control" questions. These are based on the idea proposed by Reid in 1947 and, in Backster's phrase, are intended to "capture the psychological set" of the innocent subject. Although they do not refer to the specific issue of the stabbing, they do refer to related activities in the past, to hurting someone and to lying to get out of trouble. During the pretest interview, the examiner should have rationalized these questions to the subject, suggesting that if the subject is the kind of person who has gone around hurting people and lying in the past, then it will be harder to believe in his innocence in the present instance. Such preparation will make these questions seem more important and may predispose the subject to deny them. If, as in this case, the subject says that he can answer "No" to a control question, the examiner privately assumes that this answer is a lie. Reid and Inbau's textbook explains that the examiner should select as controls only questions to which the subject shows "behavior symptoms of deception" (hesitation, breaking eye contact, squirming, etc.). Other examiners simply assume that everyone has done the sorts of things referred to in these questions and that a "No" answer *must* be deceptive.

If the subject answers "Yes" to a control question, the examiner should pretend to be surprised and concerned. He will ask, "When did you lie to get out of trouble?" and proceed to take notes of the reply. He will say, "Except for that incident, have you ever lied to get out of trouble?" and make it clear by his manner that he now expects a denial. Should the subject think of a second instance, the examiner redoubles his appearance of shocked concern, writes down that second instance, and tries again. His aim is to induce the subject to answer "No" at a point where he is at least in doubt whether this answer is strictly true. Whether the respondent is innocent or guilty of the specific offense referred to in the relevant questions, it is assumed that his answers to the control questions are actually deceptive, or at least that the subject is doubtful about whether he has answered truthfully. Therefore, this most widely used type of control question is often (somewhat optimistically) referred to as a "known-lie" control.

The scoring of the Control Question Test is straightforward. If the polygraph responses to the relevant questions are systematically larger than those elicited by the known-lie controls, the subject is considered to have been deceptive in his answers to the relevant questions. If his responses to the controls are the larger, then he is classified as truthful. The test is declared to be inconclusive if there is minimal difference in size between the two sets of responses. In the Zone of Comparison version of the CQT, advocated by Backster and illustrated by the question list shown in Table 8.1, each relevant response is compared only to the controls that are adjacent to it.

A Genuine Control Question Test

The theory of the Control Question Test is obscure and confusing, so it may be helpful to begin by indicating what the CQT clearly is *not*. Suppose that the stabbing suspect referred to earlier had committed a previous murder and that, unknown to him, we have in our possession incontrovertible evidence of his guilt. Let us call the present crime, in which he is only a suspect, Crime X and the previous one, for which he thinks that he is equally only a suspect, Crime Y. Under these unusual circumstances, the relevant question, "Did you commit Crime X?" will be psychologically equivalent to the control question, "Did you commit Crime Y?" if *the subject is also guilty of Crime X*. He should *see himself in equal jeopardy* in both cases and his "No" answers to both questions will be lies. In this imaginary situation, we would have produced a control question that truly functions as a control in the usual scientific meaning of that term. Specifically, the response to the control question provides us with a prediction of how this subject *ought to respond to* the relevant question if his denial of that question also is a lie. We are not required to assume, against common sense, that all lies are the same size or importance. We have created a situation in which we can be sure not only that the control reply is deceptive, but also that the subject, if he is guilty of both crimes, will construe both questions as about equally threatening, equally arousing. If he in fact shows much *less* physiological arousal to the relevant question, as compared to the control, then we can probably be fairly certain that he is innocent of Crime X.

What if he shows as much or more response to the relevant question? Can we then be certain he is guilty of Crime X? That is, are we justified in assuming that someone will always be more aroused by a true accusation than he will by a false charge in an equally serious matter? I find that somewhat harder to accept. The false charge might seem unfair, harder to deal with. Since the polygraph measures only relative arousal, not lying or guilt, the suspect's uncertainty and indignation about the Crime X accusation (of which he is innocent) *might* produce as much polygraphic activity as he shows to the question about Crime Y, which charge he has accepted in his mind and is prepared to "tough out."

The Control Question Test that is actually used by polygraphic examiners differs fundamentally from this genuine known-lie example. The subject's answers to the CQT's control questions are *not* "known" lies but are only assumed to be lies. Moreover, there is no attempt to match the control "lie" to the relevant lie in magnitude or importance to the subject, no attempt to predict from the control response what this person's response to the relevant question ought to be if he is lying. Therefore, the lie control question is no "control" at all in the scientific sense. It is merely a kind of comparison stimulus, just as Reid originally called it, intended to be more provocative than an irrelevant question but, for a guilty subject, *less* provocative than the relevant question. With a genuine known-lie control, a suspect who is guilty of both crimes ought to give about equal responses to the relevant and control questions—provided we have fooled him with our pretense that our evidence of his involvement in Y is no better than our evidence about X, and provided that *he* considers both crimes to be equally important or emotional (and provided that he does not bite his tongue or otherwise self-stimulate to augment his response when we ask him the control question: see Chapter 19).

Assumptions of the Control Question Test

The theory and assumptions of the CQT have never been clearly set forth by any of its proponents. Therefore, we must infer theory from practice, deduce how the test is supposed to work from the way it is administered and scored. Here is how a leading proponent describes the procedure:

> The control questions deal with acts similar to the issue under investigation but are more general in nature. They cover many years in the prior life of the subject and are deliberately vague. Almost anyone would have trouble answering them truthfully with a simple "No." ... Control questions are designed to provide the innocent suspect the opportunity to become more concerned about questions other than the relevant questions and to produce stronger physiological reactions to the control questions. If the subject shows stronger physiological reactions to the control as compared to the relevant question, the test outcome is interpreted as truthful. Stronger reactions to the relevant questions indicate deception.[2]

According to Dr. Raskin, an examiner investigating the theft of a ring might introduce the control questions as follows:

> *"Since this is a matter of a theft, I need to ask you some general questions about yourself in order to assess your basic honesty and trustworthiness. I need to make sure that you have never done anything of a similar nature in the past and that you are not the type of person who would do something like stealing that ring and then would lie about it.... So if I ask you, 'Before the age of 23, did you ever lie to get out of trouble ...?' you could answer that no, couldn't you?"* Most subjects initially answer no to the control questions. If the subject answers yes, the examiner asks for an explanation ... [and] leads the subject to believe that admissions will cause the examiner to form the opinion that the subject is dishonest and therefore guilty. This discourages admissions and maximizes the likelihood that the negative answer is untruthful. However, the manner of introducing and explaining the control questions also causes the subject to believe that deceptive answers to them will result in strong physiological reactions during the test and will lead the examiner to conclude that the subject was deceptive with respect to the relevant issues

concerning the theft. In fact, the converse is true. Stronger reactions to
the control questions will be interpreted as indicating that the subject's
denials to the relevant questions are truthful.[3]

We must be careful not to infer from this description of the technique
any unnecessary assumptions nor to set up a straw man theory that is easy
to topple but unfair to polygraphers who have faith in this method. Let us,
therefore, proceed cautiously.

ASSUMPTION 1. *A given subject will respond more strongly to a relevant
question if he answers it deceptively than if his denial is truthful. That is, if his
response would be R_I if he is innocent and R_G if he is guilty, then R_G will be larger
than R_I ($R_G > R_I$).*

This clearly is an assumption of *any* form of lie test and it seems
generally plausible. But in an area where the tradition has been to ignore
individual differences, to take a simplistic view of human nature as if all
people came off the same assembly line, it is important to emphasize that
"plausible" does not imply "inevitable." For example, Mary K. reports that
Walter, whom she met last night in a bar, drove her out into the country
and forcibly raped her. The county attorney is reluctant to prosecute
Walter unless Mary's allegations are confirmed on the polygraph. On
Mary's CQT, she is asked the relevant question, "Did you voluntarily
agree to have intercourse with Walter?" and she answers in the negative. If
Mary is lying and has accused Walter out of spite or to mollify her parents
or her husband, this question and her deceptive answer ought to constitute
an arousing stimulus. But what if Mary is telling the truth? Instead of a
consensual sex act, what if Mary's memory of this experience with Walter
are memories of fear and pain and brutal violation? How "arousing" will
this same relevant question be to her then? In Yakima County, Washington,
where all women reporting rape were required to take lie detector tests,
60% "failed" the tests due to strong reactions to the relevant questions.[4]

This first assumption of the Control Question Test may be true for
most subjects in most situations—but not for all subjects nor in all situations.

ASSUMPTION 2. *A skillful polygraph examiner formulates "known-lie" control
questions and presents them to the subject during the pretest interview in such a
way that, on the polygraph test itself, the subject will either answer these questions untruthfully or at least he will be uncertain and concerned about his
answers.*

Many polygraphers literally believe that the answers to their control questions can be assumed to be lies, for all subjects and situations. While this is absurd, it would be unfair to state Assumption 2 in this uncompromising way. Such an assumption would be patently implausible and is not a necessary or even useful component of the theory of the CQT. If all control questions *did* elicit "known lies," this would not guarantee that innocent subjects would therefore react more strongly to the control than to the relevant questions. The polygraph response is a reaction to the stimulus package that includes the question and the subject's answer. For most people, a moderately threatening control question may be more arousing if answered deceptively rather than truthfully. But this plainly does not mean that the control response will be larger than that produced by a very threatening relevant question answered truthfully. For example, if I falsely deny ever having stolen anything (the control question), I would expect to be less aroused than when I truthfully deny some serious criminal charge that may send me to prison. It is not true that all lies are psychologically equal or that any question answered deceptively will yield more polygraphic activity than any other question—no matter how relevant or threatening or emotional—that is answered truthfully. All that one can reasonably say about the known-lie control is that, properly set up and presented, it should generate more concern, stress, and arousal than an irrelevant question—and that it *may* elicit answers that the subject knows or suspects to be untrue.

This assumption begins to tax the psychological sensitivity and dramaturgical skills of the examiner because he must be able to condition (that is, deceive) the subject in specific ways. He must make the subject think that it will be detrimental to his interests to admit the sorts of misdeeds referred to in the control questions; the subject must not be allowed to catalog every lie he has ever told or every dime or pencil he has ever "stolen." And the polygrapher must make the subject think that, if the polygraph somehow shows that his answer to a control question is deceptive, then he will "fail" the test. Both of these ideas are untrue and yet the subject must be led to *think* that they are true if the control question is to produce real stress or concern.

This second assumption of the Control Question Test, like the first, may plausibly be true for many subjects, but it will not be true if the examiner is inept in enacting his role, or if the subject is knowledgeable about the test and not easily deceived, or if the subject can actually answer "No" to the control questions truthfully and without concern.

ASSUMPTION 3. *A subject's arousal response elicited by the control question* (R_C) *will be smaller than his response would be to the relevant question if he is guilty* (R_G) *but larger than his response would be if he is innocent* (R_I); *that is* $R_G > R_C > R_I$ *for all subjects.*

Here is the real nub of the theory of the CQT. There can be no argument about whether Assumption 3 is required because it is directly on this basis that the charts are scored. Notice that this third assumption includes Assumption 1, namely, that R_G will be larger than R_I. It also includes Assumption 2, namely, that R_C will be a substantial response, large enough at least to be greater than R_I. But Assumption 3 goes further, making the amazing claim that R_C will tuck neatly between R_G and R_I. The examiner has no way of knowing how strongly this particular subject will react to the relevant question if he is guilty or if he is innocent. That is, he cannot know how large this subject's R_G or his R_I would be (if he did know either of these values, he would not need a control question, but could score the chart merely by comparing the actual response with the known value of R_G or R_I). Instead, he assumes that R_G would be larger than R_I and he *also* assumes that the stimulus package of his control question, plus the subject's reply, will be *just arousing enough for this particular subject* so that it will elicit an R_C that is larger than this subject's R_I would be, while at the same time smaller than this subject's R_G would be. This truly is behavioral engineering of a precision hitherto undreamed of!

Wayne K. was fired from his responsible position at the bank because he failed a CQT based and scored on these assumptions. James Galloway served five years in an Iowa prison, convicted of murder after the jury had been told that he failed a CQT and that these tests were accurate more than 99% of the time. Although eyewitness and physical evidence fully supported his plea of self-defense, James Ray Mendoza was convicted of first-degree murder by a Wisconsin jury after it heard testimony that he had failed a CQT; he was sentenced to life in the penitentiary. This is not academic hairsplitting we are engaged in. What the CQT assumes and whether these assumptions are credible are important social questions. In the Galloway case, a typical relevant question was "Did you shoot Harry Shannon?" and I would estimate that Galloway's average response on three presentations of this question was about 6 units on an arbitrary scale of 0 to 20. The adjacent control question was "Did you ever threaten anyone with a gun?" and I would estimate his average response to this question was about 5 units. The polygrapher believed that Galloway's relevant answer was deceptive because his relevant response was (slightly) larger than his control response. But this reasoning necessarily assumes that, if Galloway was innocent and had been in Missouri at the time of the robbery as he insisted, then his relevant response would have been smaller (say, 4 units) *or* his control response would have been larger (say, 7 units): This is the concrete, real-life (and arbitrary) meaning of Assumption 3.

This brings us to an idea promulgated by Backster in 1974 under the forbidding title, the "Anticlimax Dampening Concept." Backster's notion was that

> a person's fear, anxieties and apprehensions are channeled toward the situation which holds greatest immediate threat to his self-preservation or general well-being, (and that there is) an ability within us to tune in that which may indicate trouble or danger by having our sense organs and attention set for a particular stimulus and oriented in a manner that will dampen any stimulus of lesser importance.[5]

As applied to the CQT, this hypothesis can be used to argue that the control response of a guilty subject will be "dampened," that is, made smaller than it otherwise would be, by the presence in the question list of the relevant questions, which pose a greater immediate threat to the guilty suspect's well-being. Similarly, if the subject is innocent *and if he can be made to believe that the control question poses a greater threat than the relevant question*, then his response to the control question should be larger and his relevant response should be "dampened." But Backster's hypothesis, even if it were true, would not salvage the CQT from the pits of implausibility where last we left it. Analyzing the theory of the CQT from this vantage point, we might set aside the three assumptions already discussed and substitute two others in their stead:

ASSUMPTION 1-B. *All guilty subjects will regard the relevant questions as more threatening than the control questions and therefore R_G will be larger than R_C.*

ASSUMPTION 2-B. *All innocent subjects will regard the relevant questions as less threatening than the control questions and therefore R_C will be larger than R_G.*

Again, we cannot help but be impressed by the simplistic, robotlike conception of human nature, the blithe disregard of the subtleties and idiosyncrasies of real human minds. Infallibly, the relevant question is "most threatening" to the guilty subject; therefore, we simply deny the possibility that some guilty persons might become inured to or defended against such references to their crime; we deny that any other question might be more arousing to them; we ignore the possibility that the guilty subject might covertly augment his physiological response to the control question. Universally, we are told, the relevant question is *not* "most threatening" to the innocent subject because the polygrapher will always make him believe that the control questions pose a greater threat. Therefore, the truthful rape victim will respond more strongly to the control question, "Have you ever thought of having sex with someone in the woods?" than she will to the relevant question, "Did you voluntarily agree to have sex with George Wilson last Saturday?"—even though she trembles at the mention of his name. Although James Galloway has been arrested for a robbery-murder in Davenport, Iowa, he will regard the control question, "Have you ever threatened anyone with a gun?" as more threatening to his well-being than the relevant question, "Did you shoot grocer Harry Shannon?" if he is actually innocent as he claims to be.

A psychologist must contemplate the theory of the Control Question Test with a kind of awe. The great difficulty in explaining the lie test to lawyers and laymen is that they find it hard to accept that the mythological lie detector actually teeters on such implausible premises. It seems plain that any sensible subject, whether guilty or innocent, should tend to be most aroused—and therefore to give the strongest polygraphic response—to the relevant or "Did you do it?" questions. For the CQT to work as advertised, each subject must be made to believe that the test is nearly infallible (far from true) and that giving strong control responses will jeopardize him (the opposite is true). It is implausible to suppose that all polygraphers will be able to convince all subjects of these two false propositions. Based on my analysis, one would expect that most—but not all—guilty subjects would fail the CQT—and that many innocent subjects would fail it also.

Some Real-Life Examples

How is it possible that thousands of polygraphers have managed to believe such implausible ideas? I think part of the answer lies in the fact that the theory and basic assumptions of the CQT are not usually spelled out in such bold relief. Another reason, however, is the fact that many truthful subjects actually do behave as Reid and Backster would predict! I have seen polygraph charts on which persons accused of serious crimes responded to relatively innocuous control questions with physiological reactions that were clearly stronger than the ones elicited by the relevant questions. A father accused of incest by his 14-year-old daughter, referred for testing by a California juvenile court, was diagnosed as truthful on the basis of a CQT. The daughter subsequently recanted, admitting that she was jealous of her father's multiple marriages and had wanted to punish him. A young man in New Hampshire was accused of forcible rape by his own sister, convicted, and sentenced to the state prison for 7 to 10 years. Months later, after the man had been severely beaten by his fellow inmates—convicts tend to be punitive with alleged sex criminals—he was given a lie test by Deputy Sheriff G. E. Tetreault of Exeter, New Hampshire. Observing weaker responses to questions like "Regarding your sister, did you ever have sexual intercourse with her?" than were elicited by the control questions, Deputy Tetreault concluded that the brother was innocent and initiated a further investigation. The sister, and alleged victim, was persuaded to submit to a polygraph test, failed it, and then confessed that she had made the false charge because she was pregnant by her boyfriend and had decided to sacrifice her brother in order to protect her lover. Additional corroboration was provided by the fact that both baby and boyfriend displayed a heritable physical anomaly not present in the mother's pedigree.

The murder defendant who was asked the questions shown in Table 8.1 was diagnosed as truthful by polygrapher David Raskin on the basis of that CQT. It is surprising that this defendant could have been more aroused by those innocuous-seeming control questions than he was by the relevant questions concerned with the actual stabbing. It is also surprising that Dr. Raskin put so much faith in these polygraph charts that he later asserted in print[6] that this defendant "was innocent of the murder!" without mentioning that the Canadian jury, after listening to Raskin's testimony plus all the other evidence in that case, brought in a verdict of guilty.

Based on our examination of the theory, one might expect that *most* suspects would fail the CQT whether innocent or guilty, yet it is clear that at least some persons are surprisingly unreactive to what appear to be provocative and stressful relevant questions, that some innocent suspects are able to clear their names and escape prosecution by passing the test— and perhaps a few guilty ones as well.

Our commonsense expectations are vindicated, however, by the illuminating case of Sam K.[7] Sam was on business in Phoenix in 1975 and became acquainted one evening in the motel bar with a young woman,

Mary V. According to Sam's account, the couple repaired to Mary's motel room when the bar closed, had intercourse there, smoked and talked for a time, had intercourse again, and then Sam returned to his own room and bed. Before he left, Mary wrote out for him her name and home address in pledge of future meetings and he had that scrap of paper in his wallet when he was awakened several hours later by the police, summoned to arrest him on the charge of rape. After a long and unexplained delay, Mary had called the police to make this complaint. She had also called the newspapers, the first of many actions by Mary that are uncharacteristic of most rape victims and which led the state's attorney to doubt whether he could successfully prosecute Sam K., a respectable young man with an unblemished record, on Mary's word alone. As the months went by, Mary's behavior grew increasingly eccentric. Finally, the prosecutor approached Sam's attorney with the following proposal: "Let us give your man a polygraph test. We will agree in advance to drop all charges against him if he passes. But you must similarly stipulate that we can use the test results as evidence in court if your client flunks the lie test."

Many of our states permit the introduction as evidence in court of the results of lie tests administered after such prior stipulation by both parties. In my experience, these stipulated lie test propositions are only offered by prosecutors who have a weak case. If the defendant passes, there is nothing lost anyway. But if he fails, then the weak case becomes suddenly much stronger.

Sam's attorney accepted the proposal. Sam was administered a CQT of the Zone of Comparison type—and he flunked it badly. His polygraphic reactions to the three relevant questions were strong and unequivocal, much stronger than his control responses. Now Sam was in real trouble and his attorney appealed for help. He explained to me his reason for accepting the prosecution's offer this way:

> You know, I do mostly criminal defense work and most of my clients are guilty as hell, but I really thought this guy was innocent. In fact, I had even had him polygraphed privately just to satisfy myself that he was telling me the truth. He passed that first test with no difficulty. That crazy woman kept calling the newspapers and Sam's family was upset and his employers were talking about letting him go. So when the State came to me with their deal, I thought it was an easy way to put an end to all the hassle. He was innocent, he'd already passed one lie test, all he had to do was pass another one and that would be the end of it.

Interestingly, that first private polygrapher used the old-fashioned R/I format, which one would expect to be especially hard for a reactive subject like Sam to pass. But less was at stake during that test because the results were to be confidential, for his lawyer's eyes only. And perhaps Sam regarded this first examiner, hired by his own attorney, to be friendly and sympathetic to his interests. But here we were in court, now, with the prosecution confident of a conviction. In 1975, no good studies of the validity of the CQT had yet appeared in print, and all I could offer as a witness for the defense was an explanation to the jury of the implausibility of the CQT and my opinion that it would not be at all surprising for an innocent man to respond as Sam had on such a test under these circumstances. What really saved Sam K. was a little mix-up of communication between the prosecutor and his polygrapher. The three relevant questions that this examiner employed were as follows:

1. Did you force your way into Mary V.'s motel room on the night of May 14th?
2. Did you threaten to choke Mary V. in her room on May 14th?
3. Did you rape Mary V. in her motel room on the night of May 14th?

Sam answered "No" to each of these questions and the polygraph pens gyrated wildly as he did so; there was no possibility here that another polygrapher might have scored this result as truthful or even inconclusive. But the mix-up was that Mary V.'s own account of what had happened began with her admission that she had invited Sam into her room that night after the bar closed. She said he had needed a toilet and, her room being closer, she had invited him in to use her bathroom. Sam was not accused of "forcing" his way into Mary's room that night. We know, therefore, that his answer to the first relevant question was truthful. Yet he reacted just as strongly to that question as he did to the other two.

"Can you explain, Doctor," Sam's lawyer asked me during his direct examination, "why the defendant gave such a strong reaction to that first question even though, as we know, his answer was truthful?"

> I would assume he responded as he did because this question referred to the serious crime he had been charged with, it named the woman who had brought this charge against him, and it referred to the time and place of the incident that led to his being arrested. Compared to

the so-called control questions, all three relevant questions were obviously important; all three were strong, stressful stimuli for the defendant under the circumstances and they produced strong reactions.

"Well, Doctor, given that we know the defendant's answer to the first of the three questions was truthful, have you formed an opinion as to the truthfulness of his answers to the other two questions, where his polygraph reaction was just as strong as the one he gave when we know he answered truthfully?"

> I have. I believe the only reasonable inference to be that the other two answers were as truthful as the first.

The jury agreed, and there now hangs on my wall a handsome Navaho rug, a gift from that relieved and innocent defendant after his close call. That second polygrapher had inadvertently created what I call a "Truth Control Test" by his mistake, a test format that, as we shall see later, makes considerably more psychological sense than the Control Question Test, and that was what actually saved the day for Sam K.

The Validity of the Control Question Test

The Control Question Test is regarded by most examiners as the optimum technique for specific-issue situations like those of criminal investigation, and these examples are useful for illustrating how it works—and fails to work—in real life. My commonsense analysis suggested that nearly everyone might be expected to fail the CQT, but I have listed striking instances to disprove that expectation. Polygraph examiners, in contrast, claim that the CQT almost always returns a correct diagnosis, that its assumptions, however implausible they might seem, turn out to be true over 90% of the time. Examples like the case of Sam K., which corroborate my psychological analysis, at the very least cast doubt on these claims of high validity. But the best answer would come from controlled empirical studies of CQT validity in real-life applications. For 50 years the lie detector technique gathered adherents and grew in acceptance almost entirely without benefit of such evidence. Then, in the early 1970s, there appeared several validity studies from the Reid organization that seemed to show that the CQT did indeed have surprisingly high validity. As we shall see, however, these four Reid studies actually tell us nothing about validity, but only that Reid-trained examiners, asked to read the same charts, will tend to score them the same way. But, at last, some relatively good studies of the validity of the CQT have appeared, different enough in their design so that their remarkable agreement in outcome entitles us to place at least some credence in their findings.

Horvath and Reid's[8] 40 charts were independently scored by ten polygraph examiners employed by the Reid firm. Hunter and Ash[9] used charts from ten innocent and ten guilty suspects and had them read by seven Reid polygraphers. Wicklander and Hunter[10] also used ten innocent and ten guilty charts and had them scored by six of Reid's polygraphers. Slowick and Buckley[11] reported findings based on 15 innocent and 15 guilty suspects whose charts were scored by seven polygraphers. The fatal defect shared by all four of these in-house studies is this: Instead of being estimates of lie test *validity*, they are merely demonstrations that Reid's examiners score charts in a similar way.

As we have seen, the lie test method taught by Reid is based on the assumption that deceptive subjects will give larger polygraphic responses to the relevant questions than they do to the control questions asked of them; truthful subjects are expected to show the reverse pattern. In *all* of the cases selected for independent rescoring in the four studies mentioned, the original examiner had found the charts to show these expected truthful patterns for the verified-innocent subjects, or the expected deceptive patterns for the verified-guilty suspects. Therefore, when the other Reid-trained examiners subsequently rescored these particular charts, the extent to which they agreed with the criterion was really just a measure of their agreement with the original examiner's scoring, a measure of the test's *reliability* rather than of its *validity*.

Here is how one might "prove" the validity of astrological theory by the same technique. First, we shall have one astrologer predict the sun sign (Leo, Gemini, Scorpio, and so on) of a large number of celebrities, based on their popularly known personalities and characteristics. Even if there is nothing to astrology at all, this first set of guesses will be correct some of the time just by chance. Suppose that this first astrologer makes his guesses for 1,200 celebrities and historical figures and, as chance would predict, that he is correct for 100 of them. Now let us take *just these 100 cases* and have a second astrologer predict their sun signs; suppose he is correct 80% of the time. Have we proved the validity of astrology? No, we have shown merely that the two astrologers entertain the same set of beliefs as to which personality traits go with which sun signs. We would have proved nothing at all about whether this consensus had any actual validity.

An Example of a Bad Validity Study

Recently, proponents of the lie detector have made much of the study by Raskin, Kircher, Honts, and Horowitz of CQT results selected from the files of the U.S. Secret Service.[12] Despite this study's having been completed in 1988, it has not been published in a peer-reviewed scientific journal. Instead, what the authors did must be pieced together from unpublished reports and book chapters in which the procedures and results are selectively presented.

A complication of this study that makes it difficult to compare with others is that it was not possible to verify subjects as truthful or deceptive to all of the questions they were asked on the CQT. Instead, the subjects were classified as "verified deceptive" if at least one CQT question elicited a confession and the answer to no other question could be confirmed truthful. They were "verified truthful" if an alternative suspect admitted guilt to the issue covered by a single question and the subject did not admit to guilt on the issues covered by the other questions. Hence, for many subjects, only partial verification of guilt or innocence was obtained.

From a total of 2,522 CQTs administered, 66, or about 3%, resulted in posttest confessions that verified at least one question on the test that prompted the interrogation, and 39 of these confessions verified the truthfulness of other suspects regarding their response to at least one test question. The 39 verified-truthful CQTs were administered to persons in multiple-suspect cases who were tested prior to some alternative suspect who later confessed. Thus, in this study, only about 4%(!) of all the CQTs administered were "verified" as indicating guilt or innocence, and this verification was incomplete, pertaining only to a single relevant question for many of the subjects. The representativeness of these partially verified cases is obviously open to question.

However, a more serious concern is that the method of verification virtually guaranteed that the original examiners' diagnosis would have been correct. That is, the 66 verified-deceptive CQTs were verified because the examiner diagnosed deception, interrogated on that basis, and obtained a confession. Similarly, we know that the 39 CQTs verified as truthful by the confession of an alternative suspect would have been classified as truthful by the original examiner, else he would have had little reason to test the alternative suspect.

We can therefore conclude that at least 4% of the suspects tested by the Secret Service produced CQT results like those predicted by the theory (deceptive suspects more disturbed by relevant than by control questions, truthful suspects more disturbed by the control questions). We can also conclude that other examiners, asked to score these same polygraph charts, were likely to get results similar to those obtained by the original examiners. But these results are entirely compatible with the assumption that there is no relationship *whatever* between the veracity of the suspect and his score on the CQT!

Let us suppose that 50% (1,261) of the 2,522 suspects tested were in fact guilty. Let us assume further that the CQT identified deception in these individuals with only chance accuracy. This would result, by chance alone, in 630 suspects being classified as deceptive. All or most of these 630 would be interrogated and, as in this case, some 66 of them might confess their guilt. Since some of these 66 would be involved in cases with multiple suspects, their confessions would exculpate the alternative suspects. These alternative suspects were most likely tested and diagnosed truthful prior to the testing of the guilty suspect who confessed, otherwise the suspect who confessed would most likely not have been tested because the guilty party would already have been identified. Note that under these circumstances, the testing of these subjects would be correct 100% of the time even though the test itself has only chance accuracy. If we now take the charts from these cases and have them blindly rescored, then, because scoring is reliable, we are likely to obtain nearly the same results that the original examiners obtained. However, these results, suggesting near infallibility, are totally misleading and tell us nothing about CQT accuracy.

Raskin and his colleagues emphasize a unique feature of their study as though it represents a significant methodological refinement over other reports. In addition to requiring a confession to substantiate guilt, they also required the presence of "independent corroboration" of the confession in the form of some type of physical evidence. Although this requirement would appear to eliminate the occasional false confession, it does not deal with the fundamental problem inherent to the use of confessions to establish ground truth. The problem with this requirement is that the corroborating physical evidence is not independent of test outcome. Had the suspect not failed the CQT, there would be no confession, and had there been no confession, there would have been no opportunity to recover the physical evidence. Hence, there is nothing "independent" about the corroborating evidence. Just like the confession, it too is dependent on the suspect having failed the polygraph test.

In short, although much has been made of the Secret Service study, as if it had demonstrated a high degree of accuracy for the CQT, when properly analyzed it can be seen to be wholly without probative value. The fact that this study has not been accepted for publication in a peer-reviewed scientific journal illustrates the utility of impartial peer review as a minimum criterion for consideration of scientific claims.

CQT Studies Published in Scientific Journals

Laboratory Studies   The studies that have achieved publication, although none of them meets the criteria set out earlier for an adequate validity assessment, do permit certain limited conclusions to be drawn. First, there are a number of studies in which volunteer subjects are required to commit a mock crime and then to lie about it during a CQT examination.[13] Control subjects do not commit the crime and are truthful on the CQT.[14] Instead of fear that failing the CQT will lead to punishment (such as criminal prosecution), subjects in these studies were motivated by a promise of a money prize if they were able to be classified as truthful on the CQT.[15] In these highly artificial circumstances, CQT scores successfully discriminated between the two groups with an accuracy of about 90%.

When the circumstances are made somewhat more realistic, however, even this mock crime design produces results similar to those reported in the better field studies (discussed below). Patrick and Iacono, for example, using prison inmate volunteers, led their subjects to suppose that their failing the CQT might result in the loss to the entire group of a promised reward and thus incur the enmity of their potentially violent and dangerous comrades.[16] Under these circumstances, nearly half of the truthful subjects were classified erroneously as deceptive.[17] In another study, Forman and McCauley permitted their volunteer subjects to choose for themselves whether to be guilty and deceptive or innocent and truthful.[18] Those who elected to be truthful knew that their reward would be smaller but presumably more certain.[19] This manipulation is analogous to crime situations where an individual is confronted with an opportunity to commit a crime with little likelihood of getting caught (e.g., an unlocked car with a valuable item in sight, a poorly watched-over purse or briefcase, etc.) and must decide whether to take advantage of the opportunity. By thus increasing the realism of the test conditions, Forman and McCauley probably also obtained a more realistic result, with about half of their truthful subjects being erroneously classified as deceptive.

Thus, although mock crime studies with volunteer subjects clearly do not permit any confident extrapolation to the real-life conditions of criminal investigation, it does appear that the designs with the greater verisimilitude, which threaten punishment or which merely permit subjects to decide for themselves whether to be truthful or deceptive, demonstrate that the CQT identifies truthful responding with only chance accuracy.

Field Studies   There are four field studies of the CQT that have been published in scientific journals.[20] Three of the four include cases where the verification of guilt and innocence was not entirely dependent on polygraph-induced confessions. Although not published in a peer-reviewed journal, the Barland and Raskin study (Barland's Ph.D. research under Raskin's direction),[21] which its authors have repudiated, produced two interesting results. First, the principal scientific advocate of the CQT, Dr. Raskin, who independently scored all the charts, classified more than half of the innocent suspects as deceptive. Second, knowing that most of the suspects tested were probably guilty, Raskin scored 88% of them as deceptive. Since 78% of them were in fact guilty, if we classified 88% of the total group as deceptive and the rest as truthful *entirely at random*, we should achieve an average accuracy of 71%. Raskin's average accuracy, based on the polygraph charts, was also 71%.

The studies by Horvath and by Kleinmuntz and Szucko both used confession-verified CQT charts obtained respectively from a police agency and the Reid polygraph firm in Chicago. The original examiners in these cases, all of whom used the Reid clinical lie test technique, did not rely only on the polygraph results in reaching their diagnoses but also employed the case facts and their clinical appraisal of the subject's behavior during testing. Therefore, some suspects who failed the CQT and confessed were likely to have been judged deceptive and interrogated based primarily on the case facts and their demeanor during the polygraph examination, leaving open the possibility that their charts may or may not by themselves have indicated deception. Moreover, some other suspects were cleared by confessions of others, even though the cleared suspects, judged truthful using global criteria, could have produced charts indicative of deception. That is, the original examiners in these cases were led to doubt these suspects' guilt in part regardless of the evidence in the charts and proceeded to interrogate an alternative suspect in the same case who thereupon confessed. For these reasons, some undetermined number of the confessions that were criterial in these two studies were likely to be relatively independent of the polygraph results, revealing some of the guilty suspects who "failed" it. The hit rates obtained in these studies are indicated in Table 8.2.

In the study by Patrick and Iacono, 13 of the 20 innocent suspects were confirmed as such independently of polygraph results (e.g., the complainant later discovered the mislaid item originally thought to have been stolen). As can be seen in Table 8.2, 9, or 45%, of these 20 innocent suspects were wrongly classified as deceptive by the CQT. Only one guilty suspect could be confirmed as such from file data independent of CQT-induced confessions; his charts were classified as inconclusive by the CQT. The remaining guilty suspects in the Patrick and Iacono study were all classified solely on the basis of having been scored as deceptive on the polygraph and then interrogated to produce a confession. Understandably, when examiners trained in the same method of scoring independently rescored these charts, they agreed with the original examiners in 98% of cases.

The recent study by Honts illustrates that publication in a refereed journal is no guarantee of scientific respectability. The meticulous study by Patrick and Iacono was done with the cooperation of the Royal Canadian Mounted Police (RCMP) in Vancouver, B.C., and showed that nearly half of the suspects later shown to be innocent were diagnosed as deceptive by the RCMP polygraphers. This prompted the Canadian Police College to contract with Honts, one of the Raskin group, to conduct another study. A

Table 8.2. Summary of Studies of Lie Test Validity That Were Published in Scientific Journals and That Used Confessions to Establish Ground Truth[a]

	Horvath[b] (1977)	Kleinmuntz & Szucko (1984)	Patrick & Iacono (1991)	Honts (1996)	Mean
Guilty correctly classified	21.6/28 77%	38/50 76%	48/49 98%	7/7 100%	114.6/134 85.5%
Innocent correctly classified	14.3/28 51%	32/50 64%	11/20 55%	5/5 100%	62.3/103 60.5%
Mean of above	64%	70%	77%	100%	73.0%

[a]See note 20, Chapter 8 notes, for names of studies used in this table.
[b]Horvath had each subject's chart read independently by 10 polygraphers; hence, e.g., of 280 evaluations of the charts of the 28 innocent suspects, 143 were classified as truthful.

polygraphy instructor at the college sent Honts charts from tests administered to seven suspects who had confessed after failing the CQT and also charts of six suspects confirmed to be innocent by these confessions of alternative suspects in the same crimes. Knowing which were which, Honts then proceeded to rescore the charts, using the same scoring rules employed by the RCMP examiners. Those original examiners had, of course, scored all seven guilty suspects as deceptive; that was why they proceeded to interrogate them and obtained the criterial confessions. Using the same scoring rules (and also knowing which suspects were in fact guilty), Honts of course managed to score all seven as deceptive also. The RCMP examiners had scored four of the six innocent suspects as truthful and two as inconclusive. We can be confident that all innocent suspects classified as deceptive were never discovered to have been innocent because, in such cases, alternative suspects would not have been tested, excluding any possibility that the truly guilty suspect might have failed, been interrogated, and confessed. Honts, using the same scoring rules and perhaps aided by his foreknowledge of which suspects were innocent, managed to improve on the original examiners, scoring five of the six as truthful and only one as inconclusive. The difference in Honts's findings from those of the other studies summarized in Table 8.2 is striking.

Surely no sensible reader can imagine that these alleged "findings" of the Honts study add anything at all to the sum of human knowledge about the true accuracy of the CQT. How it came about that scientific peer review managed to allow this report to be published in an archival scientific journal is a mystery. Since the author, Honts, and the editor of the journal, Garvin Chastain, are colleagues in the psychology department of Boise State University, it is a mystery they might be able to solve.

Verdict

The Control Question Test, widely regarded among polygraphers as their most refined technique, is the only lie detection method to have been seriously studied with respect to its validity. As we have seen, some of these studies are defective or irrelevant; none of them are definitive. Because of the contamination resulting from reliance on polygraph-induced confessions as criteria of ground truth, the CQT's accuracy, especially in detecting guilty suspects, is overestimated by these studies to an unknown extent. As we shall see in Chapter 18, naive subjects can learn to

beat the CQT with less than an hour's instruction, using covert countermeasures that experienced polygraphers cannot detect. The burden of
proof, however, is (or should be) on the proponents of the method. Can
they substantiate their claims of near-perfect accuracy? After listening to
a week of testimony from critics of polygraphy and from such leading
polygraphers as John Reid and Richard Arther, Justice Morand concluded:

> The polygraph examiners had many opportunities to answer the prob-
> lems and criticisms suggested by psychologists and physiologists.
> Unfortunately, their response was invariably that the criticisms were
> not valid because, in their experience, the test worked. I have come to
> the conclusion that I must accept the evidence of the psychologists and
> physiologists, which is consistent with both my common sense and my
> personal experience, that all individuals do not react in identical ways
> in a given situation, and that programming human responses is at best
> imperfect. In my opinion there is a real possibility that many innocent
> persons accused of crime would be unconcerned with what has been
> suggested to me are good control questions in comparison with the
> actual accusation. I have no doubt that some people do react as poly-
> graph operators insist they must, but I am not convinced that this latter
> group of people would be an overwhelming proportion of our popu-
> lation.[22]

Chapter 9

THREE DIFFERENT
VERSIONS OF THE CQT

There is no worse lie than the truth misunderstood.

—William James

Man is the only animal that blushes—or needs to.

—Mark Twain

The Directed Lie Test (DLT)

The Directed Lie Test is a form of the CQT in which the subject is instructed to answer each control question deceptively.[1] "You've told a lie sometime in the past, haven't you? Well, I'm going to ask you about that on the test and I want you to answer 'No.' Then you and I will both know that answer was a lie and the tracings on the polygraph will show me how you react when you're lying." The DLT is scored in the same way as a standard CQT. Advocates believe that the DLT is an improvement because there is greater certainty that the subject's answers to control questions are false. Moreover, the DLT can be more nearly standardized than the CQT because the same control questions can be used in every case since the subject does not have to be duped into giving what he suspects may be a false answer.

For example, in the Horowitz *et al.* study described below, the three directed lie controls were as follows:

> Have you ever told a lie?
> Have you ever broken a rule or regulation?
> Have you ever made a mistake?

The essential differences between the DLT and the standard CQT are as follows:

1. For the CQT, the examiner assumes the control answer is deceptive, whereas, for the DLT, both the examiner and the suspect know that these answers are false.
2. The subject answers the control questions deceptively on instruction rather than by choice.
3. The DLT, like one version of the stim test described earlier, gives the subject the (false) impression that his polygraph reactions to the directed lie questions provide the examiner with a kind of criterion template of how he personally reacts when lying. Although deliberately misleading, this impression might persuade some subjects that the procedure has a scientific basis.

Assumptions of the Directed Lie Test   The Directed Lie Test assumes that an innocent suspect will be more disturbed when instructed to answer falsely about some past misdeed than while truthfully denying an accusation concerning the crime of which he stands accused. Just as for the CQT, the "control" questions are not controls in the scientific sense, and there is no discernible reason for supposing that innocent persons will be reliably more disturbed while denying on instruction some universal human failing than while truthfully denying some recent and serious misdeed that poses a genuine threat.

Proponents of the conventional CQT argue that one of its strengths derives from the fact that control questions are never identified as such to subjects.[2] Indeed, subjects are deliberately misled to believe that lying to the control questions will generate a "deceptive" verdict. To the extent that this deception is successful, an advantage of this approach is that unsophisticated guilty subjects may not figure out that it is to their advantage to try to augment their responses to these questions. However, it is obvious that even unsophisticated guilty suspects would be able to identify and understand the significance of the directed lie questions. They could easily self-stimulate (e.g., bite their tongues; see Chapter 19) after each directed lie answer in order to augment reactions to these "control" questions and thus defeat the test.

Horowitz et al.[3] conducted the only study of the DLT to be reported in a scientific journal. This was a laboratory study in which half of a group of volunteer subjects were required to commit a mock crime and then to deny it in the polygraph examination. The other or "innocent" subjects did not participate in the mock crime and were instructed to be truthful on the lie test. Some subjects were given a standard CQT and others were given the directed lie format. None of the subjects, of course, were told to employ countermeasures to beat the test. It was found that 80% of the volunteers tested with the DLT were correctly classified as to "guilt" or "innocence," compared to only 67% of the subjects tested with the CQT. The latter result is substantially lower than in previous reports by the Raskin group of CQT accuracy in mock crime studies.[4] But, because these *are* mock crime studies, with no penalty for "failing" the lie test, they do not tell us anything that can safely be extrapolated to the real-life situation of criminal interrogation.

In evaluating these various techniques, it is useful to imagine oneself as the subject. Suppose you are a volunteer in a laboratory experiment and know nothing about the mock crime that other subjects have been obliged to walk through; you have been assigned to the "innocent" group and told to truthfully deny the crime to the polygrapher. Hooked up to the polygraph, you are asked: "Did you take the diamond ring from the desk in the office?" and you answer, truthfully, "No." Then you are asked: "Have you ever told a lie?" and, as previously directed, you answer that one, also, "No." Which of those two questions would you expect to elicit the larger ripple of involuntary response from you, the stronger "internal blush"? If the polygrapher concludes wrongly that you had been assigned to the "guilty" group, that's his problem, nothing to do with you; you will not be prosecuted or punished. But to deny a universal human failing, even on instruction, is at least a little bit embarrassing. Now let us change the scene to the police station where you have been taken, charged with having sexually abused your grandchild.[5] Once again you are innocent but this time something real and alarming is at stake. Which of those two questions will make the polygraph pens flutter the most in this different context?

Verdict   It is surprising that the Raskin group, perhaps the only people with scientific training who are currently practicing as professional polygraphers, has come to the conclusion that the DLT is a significant advance over the CQT. But, then, it is surprising that people with scientific training would have been willing to claim on the basis of existing evidence that either technique is at least 90% accurate in diagnosing both truth and falsehood in real-life criminal interrogation. The Directed Lie Test, like the CQT, is implausible on its face and unsupported by competent research.[6]

The Positive Control Test (PCT)

Another question format used by some polygraphers involves using the relevant question itself as its own control.[7] Each relevant question is asked twice and the subject is instructed to answer truthfully on one presentation and falsely on the other. This is an ingenious variation that appears to circumvent some of the difficulties involved in the CQT. Here, the identical question is associated once with a lie and once with a truthful answer. Therefore, it is argued, the only factor that should influence the size of the physiological response—the only factor that could account for one response being greater than the other—is that one of the answers is deceptive and the other is not.

The PCT also has the great advantage that it can be used not only in specific-issue situations, as arise in criminal investigation, but also in the once-profitable business of employment screening, where one has a whole series of unrelated relevant questions to ask but where the sorts of control questions used either in the CQT or the DLT are not feasible.

Assumptions of the Positive Control Test   To get a feel for the PCT in action, let us imagine a situation in which Mary X. has accused John Z. of forcible rape. Both John and Mary acknowledge that sexual intercourse occurred but John claims it was consensual, whereas Mary insists that she was threatened and forced against her will. The district attorney tells Mary that he will not prosecute unless she can corroborate her accusation by passing a lie detector test. For simplicity, we shall assume that the PCT employed will involve only two questions, each asked twice. Mary is instructed to lie the first time each question is presented and to answer the repetition truthfully.

1. Did you voluntarily agree to have intercourse with John?
2. Did John use threats or force you to have intercourse with him?

Suppose first that Mary *was* threatened, forced, and criminally assaulted, just as she contends. Then her second answers—her spontaneous answers—will be true, while her first answers—the forced answers—will be false (not lies, since she is not trying to deceive anyone, but merely untrue answers given because the examiner requires her to give them). According to the theory of the PCT, her polygrams will show stronger reactions to the forced answers, because they are untrue, than to the spontaneous answers, which are truthful. Is this a reasonable expectation? When Mary asserts "Yes" (meaning "Yes, I was raped!") and "No" ("No, I was not willing!"), we must expect these answers to be associated with considerable emotion and arousal. When, as required by the test, she gives the opposite answers, can we safely assume that she will be still more aroused? She might—or she might not. I doubt that any prudent psychologist would want to hazard a prediction.

Now suppose that Mary's accusation is false, that her affair with John was consensual but that, when her husband learned about it, Mary decided to protect herself by charging that she had been forced. On this assumption, Mary's "spontaneous" answers will be real lies and her "forced" answers will be truthful. The PCT requires us to expect that her pattern of polygraphic reaction will now be reversed, larger responses associated with the spontaneous answers rather than with the forced answers. Certainly those spontaneous answers, those false allegations that might send John to prison, should be accompanied by emotional arousal. But if one has made a false charge, and then, while still connected to the polygraph, one is required to utter the truth (which one hopes will not be believed), surely that utterance also will involve an emotional reaction. Which response will be the stronger? Again, it is impossible to predict. One must expect some Marys to react one way and some Marys the other.

As was true for the CQT and the DLT, analysis of the Positive Control Test is difficult because none of these tests in fact provides an actual "control" at all—a basis for predicting how *this* subject should respond to the critical stimulus (the relevant question, in the case of the CQT and the DLT, or the spontaneous answer, in the case of the PCT), whether he is being truthful or deceptive. A false answer that one is required to give does not provide a good prediction of how one should react when spontaneously lying. A truthful answer that one is required to make, hoping that it will not be believed, cannot predict the arousal one might show when spontaneously telling the truth. It is not true that, in the PCT, the identical stimulus is associated with both a deceptive and a truthful answer. The identical *question* is associated with both answers, but the total stimulus, the whole set of conditions that influences the respondent's emotional reaction, includes the question, the respondent's knowledge of the real truth, *and* whether the truthful answer is forced or given spontaneously. The examiner does not know if he is dealing with a spontaneous lie and a forced truth or with the converse—that is why he is giving the test—and the polygrams cannot be relied upon to tell him.

ASSUMPTION 1. *People in general will be more aroused when uttering a spontaneous lie than when, by instruction, they are answering the same question truthfully (although they contend that truthful answer to be untrue).*

ASSUMPTION 2. *People in general will be more aroused when, on instruction, they are giving a false answer than they will when answering the same question spontaneously and truthfully.*

The appropriate reaction to both of these assumptions is simply to ask, "Why?" They seem arbitrary and capricious. No accepted principle of psychological doctrine supports them, and common sense, often a more dependable guide, rejects them. Like so much of lie detector theory, they imply an exceedingly simplistic view of human nature, a knee-jerk uniformity of reaction from one person or situation to another, a uniformity that is belied by ordinary observation and confuted by many years of psychological research. Those who would urge such assumptions must accept a heavy burden of proof. Experiments that appeared to confirm such propositions would cause a sensation among psychologists and would be subjected to searching criticism because all previous experience has indicated that, in such complex situations, human behavior is variable and inconsistent.

One reason why the PCT is preferred by some polygraphers is that it is especially effective at doing what any polygraph test does best—eliciting confessions. When a deceptive subject is told that he must lie the first time he answers the question, he is frequently confused: "Let's see, 'Yes' is the truth, and 'No' is what I want them to believe, but now he's told me to lie, so should I say 'Yes' or 'No'?" Getting the respondent confused is a common tactic of interrogation; he may stumble and reveal himself and then give up altogether.

Verdict   The only published empirical study of the validity of the Positive Control Test I have been able to find is that by Forman and McCauley.[8] This was a laboratory mock crime study in which subjects were allowed to choose for themselves whether to be "guilty" and then have to lie on the test, or to be "innocent" and then be truthful on the test (and also to receive a smaller reward for participation than the "guiltys" received). Each subject was given a short PCT, a CQT, and a three-item GKT as well, and all three techniques produced accuracies in the range from 68% to 78%. Thus, we must conclude that the assumptions of the PCT are not strongly supported either by external evidence or by intrinsic plausibility.

The Truth Control Test (TCT)

Reid's 1947 proposal for the use of a "guilt complex" comparison question never really caught on among polygraphers. Although the name "guilt complex" is unfortunate, with its unnecessary psychiatric connotations, Reid explained the principle with admirable clarity:

> The "guilt complex" question is based upon an entirely fictitious crime of the same type as the actual crime under investigation, but one which is made to appear very realistic to the subject.... The purpose of the "guilt complex" or fictitious crime question is to determine if the subject, although innocent, is unduly apprehensive because of the fact that he is suspected and interrogated about the crime under investigation. A reaction to the fictitious crime question which is greater than or about the same as that to the actual crime question would be indicative of truth-telling and innocence respecting the real offense. On the other hand, however, a response to the actual crime questions, coupled with the absence of a response to the fictitious crime question, or by one considerably less than that to the actual crime questions, would be strongly indicative of lying regarding the offense under investigation.[9]

For example, when James Galloway was interrogated about that robbery-murder in Davenport, Iowa, he might also have been questioned about another holdup in Des Moines in which the victim also died. "This happened just the day before the Davenport killing, Galloway, and we have witnesses who identified your picture in both cases." But, unknown to the suspect, there was no such killing in Des Moines on the date in question so we know that his denial of that crime is truthful. Here we have the obverse of the genuine known-lie situation discussed earlier: Crime X, of which our subject is a suspect, and Crime Y, of which we know he is innocent because Crime Y never happened. If our man is also innocent of X, then it might not be too difficult to make him think that he is in equal jeopardy with respect to both crimes; they are equally serious, the evidence implicating him is similar in both cases, and in both cases he knows that he was somewhere else. This fictitious crime or "known-truth" question, then, will provide an estimate of how this subject ought to respond to the similar question about Crime X (if he is innocent); that is, the known-truth question will be a genuine control stimulus.

The test format would be like the CQT except that the three known-lie control questions would be omitted and replaced by three known-truth controls, three questions about Crime Y that essentially parallel the three relevant questions asked about Crime X. Similar polygraphic responses to the relevant and control questions would lead to a diagnosis of truthful. Much larger responses to the relevant than to the control questions would be scored as deceptive. The test administered to Sam K. in the example discussed in Chapter 8 could be interpreted as a TCT in which what was intended as the first relevant question was actually a known-truth control. Because his reaction to this control question was similar to the responses produced by the two actual relevant questions, we (and Sam's jury) inferred that Sam probably answered the relevant questions truthfully also.

Assumptions of the Truth Control Test   As far as I can determine, a test based strictly on the TCT format has never been administered. Polygraphers sometimes include a single "guilt complex" question in the context of a CQT, but without any effort to prepare the subject so that he will regard the two crimes as equally real and equally threatening. It is obvious that the required deception of the subject would not always be possible, but with a little ingenuity and planning, the more plausible TCT could frequently replace the CQT. Consider, for example, the murder suspect who was asked the known-lie test questions given in Table 8.1. After he had agreed to take a polygraph test, he might have been called back to be interrogated, not by the polygrapher but by the detective who had questioned him earlier about the real crime.

> "Where were you on June 4th of last year?"
> "I don't know. In town, I suppose."
> "Did you know a man named Lee Wong who lived on McNair Street?"
> "No, I never heard of a Lee Wong. Why?"

"Lee Wong was stabbed in his apartment last June 4th. We have wit-nesses who described a car like yours being parked in back of that apartment. And they have identified your picture as the man who they saw running out of that apartment about 10 p.m. on the night of June 4th."

"Well, they're wrong! I've never been to any apartment on McNair and I don't know any Lee Wong. Your witnesses are lying!"

A day or two later when the polygraph test is administered, this defendant should have two crimes on his mind. If he is innocent of the real one, Crime X, then he ought to be feeling equally victimized about both of them. If he is now given a Truth Control Test like the one shown in Table 9.1, his polygraph reactions to the three control questions might provide reasonable estimates of what his responses should be to the relevant questions if he is also innocent of Crime X.

ASSUMPTION 1. *The examiner, with help from the investigating officers, will be able to persuade the subject that he is in equal danger of prosecution for a fictitious crime as serious as the one for which he is a real suspect.*

This deception would take more planning and stage-managing than the deceptions involved in the usual CQT. One would have to be sure that the subject does not have an ironclad alibi for the fictitious crime and that his attorney does not have time to discover that no such crime is actually

Table 9.1. Example of a Truth Control Test

1.	Were you born in Hong Kong?	Yes	(Irrelevant)
2.	In the crowd on Dumfries Street on January 23, 1976, did you cut anyone with a knife?	No	(Relevant)
3.	In the apartment house on McNair Street on June 4, 1975, did you cut anyone with a knife?	No	(Truth control)
4.	Is your first name William?	Yes	(Irrelevant)
5.	On June 4th, 1975, did you stab Lee Wong?	No	(Truth control)
6.	On January 23, 1976, did you stab Ken Chiu?	No	(Relevant)
7.	Apart from the two cases we have talked about, have you ever stabbed anyone?	No	(Outside issue)
8.	Do you smoke cigarettes regularly?	No	(Irrelevant)
9.	Do you actually know who killed Ken Chiu?	No	(Relevant)
10.	Do you actually know who killed Lee Wong?	No	(Truth control)

Note. The suspect in the murder of Ken Chiu has also been questioned about an earlier stabbing of one Lee Wong and told that eyewitnesses have identified him as being involved in both crimes. The murder of Lee Wong is a fiction, but this subject has been persuaded that it really happened.

under investigation. There are obvious ethical questions to be raised about deceiving a subject in this way; I return to these later.

ASSUMPTION 2. *An innocent suspect will regard both accusations as posing a similar threat to his well-being and will therefore be about equally aroused by the relevant and control questions.*

This assumption again hinges on the skillfulness of the deception. If the arousal elicited by the control questions is not clearly greater than that produced by the irrelevant questions, the TCT should probably be scored as inconclusive.

ASSUMPTION 3. *A subject who is guilty of the actual crime should be more disturbed by the relevant than by the control questions. Therefore, a polygraph chart showing small irrelevant responses, moderate control responses, and large relevant responses will indicate that the subject is lying about the actual crime.*

Since the polygraph registers merely physiological arousal, not lying, any form of lie test is inferential and subject to error. This deceptive pattern of strong relevant responses with moderate control responses might also be shown by a truthful subject who simply has not been sufficiently convinced of the alleged danger posed by the fictitious crime. Similarly, the truthful pattern of equal relevant and control responses might be shown by a deceptive subject who is guilty and lying about Crime X but has been unnerved and bothered by the false accusations relating to Crime Y. Like the CQT, the TCT is predicated on the successful deception of the subject, so both tests become invalidated when used with sophisticated subjects. But the known-truth control question has at least the advantage of being a genuine control stimulus designed to estimate how this subject should respond to the relevant question if he is innocent. It is unfortunate that no research has been done on the Truth Control Test, not even in the laboratory situation, where at least the relative merits of different test formats can be usefully compared.

A study by Podlesny and Raskin used a single guilt complex question in the context of a conventional CQT but there was no effort to make the "real" and the "fictitious" crimes equal in significance. The innocent subjects heard a description of the theft of a ring, the mock crime that the guilty subjects enacted, but were merely told that they would also be asked about the theft of a watch, the subject of the guilt complex question. Of the guilty subjects tested with the "stolen" ring in their pockets, 85% reacted more strongly to relevant than to the known-truth questions, while, as expected, the innocent subjects reacted about equally to the relevant and guilt complex questions.[10]

Verdict   The Truth Control Test would appear to be the most plausible approach to the polygraphic detection of deception. The TCT is not used by professional polygraphers at the present time, except for the occasional halfhearted inclusion of a guilt complex question as part of a CQT. There is no doubt that the TCT could not be used in many situations and it would always require considerable ingenuity and planning to establish the needed deception. The greatest weakness of the Truth Control Test is the impractical and possibly unethical deception that must be perpetrated on the person tested. Moreover, of course, like the CQT and the DLT, which also depend on the examiner's being able to deceive the subject, the TCT would not be likely to work for long before word got around within the criminal community that the other crime never happened.

The Peak of Tension Test (POT)

The Peak of Tension Test was one of Leonarde Keeler's inventions. Suppose a kidnapped child has been found slain and a suspect is arrested. The bereaved parents have a good photograph of the child taken shortly before he was abducted. Similar photographs of six other children of about the same age are obtained by the police. The actual kidnapper would have no difficulty picking out the picture of his victim from the group of seven, but an innocent suspect would be equally likely to choose any one of the group. The correct picture is placed in, say, position 5 in a sequence and the photographs are shown to the suspect, one at a time, while he is connected to the polygraph. Under the circumstances, he may respond to each picture, especially to the first one in the series, but only if he recognizes the murdered child should he show a much stronger reaction selectively to picture 5. Keeler and his followers also looked in the polygraph record for evidence of mounting tension as the examiner proceeded through the series of pictures, with a peak of tension at picture 5 and a relaxation thereafter. The POT is the prototype of a fundamentally different method of polygraphic interrogation designed not to detect lying but to detect *guilty knowledge*. The POT, or Guilty Knowledge Test, depends on the investigator's having knowledge of details of the crime that the suspect should recognize only if he is guilty—if he was there and possesses guilty knowledge. I consider this promising technique in detail in Part IV of this book.

A criminal interrogation can be thought of as involving three actors: the examiner (E), the guilty suspect (G), and the innocent suspect (I). As we have said, the POT or Guilty Knowledge Test requires one or more pieces of information to be shared by E and G but not by I. In some rare situations a useful test could be based on information shared by E and I but not by G. A suspect claims to have been somewhere else at the time of the crime, say, in his college classroom. By interviewing other students in that class, E may be able to compile a list of events or topics discussed that should be known to the suspect if his alibi is truthful, but not if he cut his class that day in order to rob the bank. If the information in question is known to both G and I but not to E (for example, "Did you do it?"), then we have the familiar lie detection situation. But when some specific item of information is known only to G and neither to I nor to E, then the polygrapher may decide to use the Searching Peak of Tension Test, or SPOT.

Assumptions of the SPOT   Employing the kidnapping example once again, suppose a likely suspect is in hand but the victim has not yet been found, alive or dead. A large map of the area is set up before the suspect while he is connected to the polygraph. The map is marked off into quadrants labeled A through D and the suspect is questioned thus: "Where is the child now, Ed? Is she in section A? Is she in section B?..." After several repetitions, if one quadrant seems to elicit greater poly-graphic activity than the others, an enlarged map of that area can be substituted, similarly divided into sections. With patience, luck, and a responsive subject, such a procedure has been known to lead to the loca-tion of a hidden grave, a cache of stolen property, or a discarded murder weapon. This is not a lie detection method, but purely a guide to investiga-tion. When it works, it provides its own verification.

The SPOT is frequently used in another, less interesting and also less legitimate fashion. A store is experiencing losses and decides to have polygraph tests administered to its employees. In addition to several charts based on R/I or CQT questions, the examiner asks a series of questions like this:

> "How much of these losses are you responsible for, Mr. Jones? Would $100 cover the amount you have taken? Would $200 cover it? $500? $1,000? $2,000? ..."

I have seen one real-life example that ran from $100 all the way up to $150,000. What that examiner had in mind was the notion that, if his subject responded especially strongly to some figure in that broad span, this would prove that he *was* responsible for some of the losses and also pinpoint how much he had taken. Suppose such a list of figures was repeated several times, the amounts being ordered differently each time, and suppose that the subject does plainly react each time to one figure, say, $2,000, more than to the others. What can we conclude from that? One possibility is that the subject has been stealing from the store and that $2,000 seems to him to be the approximate total of his peculations. Other possibilities are that $2,000 is the amount he has in the bank or owes on his car, or that the examiner seemed to pause after $2,000 the first time through the list and this marked that figure in a special way. If such a pattern of responding was observed, one could not blame the examiner for inter- rogating the subject about it:

"You always respond to $2,000, Mr. Jones. How do you explain that?"

But it would be unfair and unreasonable to punish Mr. Jones, to fire him or discipline him in some way, merely on the basis of this sort of ambiguous evidence.

Verdict   The Searching Peak of Tension Test is its own justification when it leads to the discovery of useful physical evidence or elicits a valid confession. The mere occurrence of consistent responding to some item in the series, by itself, is hopelessly ambiguous and provides no legitimate basis for any conclusions about the veracity of the subject.

Chapter 10

POLYGRAPH SCREENING TECHNIQUES

It is obviously a most effective protection for legitimate secrets that it should be universally understood and expected that those who ask questions which they have no right to ask will have lies told to them.

—H. SIDGWICK, *The Methods of Ethics*

If an investigator screened a group comprising 50% liars by flipping a coin, diagnosing a "liar" every time it came down heads, he would detect the actual liars in the group with an accuracy not far short of the polygraph.

—M. PHILLIPS, MRCP, A. S. BRETT, M.D., & J. F. BEARY III, M.D.[1]

Prior to the Employee Polygraph Protection Act of 1988 (Figure 10.1), employee screening was by far the most common application of the polygraph, and because the Act curiously exempted federal agencies from its proscriptions, federal police and security agencies continue to require that job applicants submit to this procedure. A valuable insight into this screening technique as it is actually practiced was provided by a book, *Preemployment Polygraphy*, published in 1984 by two well-known members of the polygraph fraternity, Robert J. Ferguson and Chris Gugas, Sr. I had the pleasure of reviewing that text for the American Psychology Association's

U.S. DEPARTMENT OF LABOR

EMPLOYMENT STANDARDS ADMINISTRATION

Wage and Hour Division
Washington, D.C. 20210

NOTICE

EMPLOYEE POLYGRAPH PROTECTION ACT

The Employee Polygraph Protection Act prohibits most private employers from using lie detector tests either for pre-employment screening or during the course of employment.

PROHIBITIONS

Employers are generally prohibited from requiring or requesting any employee or job applicant to take a lie detector test, and from discharging, disciplining, or discriminating against an employee or prospective employee for refusing to take a test or for exercising other rights under the Act.

EXEMPTIONS*

Federal, State and local governments are not affected by the law. Also, the law does not apply to tests given by the Federal Government to certain private individuals engaged in national security-related activities.

The Act permits *polygraph* (a kind of lie detector) tests to be administered in the private sector, subject to restrictions, to certain prospective employees of security service firms (armored car, alarm, and guard), and of pharmaceutical manufacturers, distributors and dispensers.

The Act also permits polygraph testing, subject to restrictions, of certain employees of private firms who are reasonably suspected of involvement in a workplace incident (theft, embezzlement, etc.) that resulted in economic loss to the employer.

Figure 10.1. The federal Employee Polygraph Protection Act prevents most private-sector job applicants from having to submit to lie detector testing. But, if your employer is investigating a loss, or if you are applying for a job with the government, you may find yourself subjected to this "fourth degree." The current Act needs strengthening.

journal of book reviews, *Contemporary Psychology*, and I can do no better than to reproduce the text of that review below.

"... In a Tumor in the Brain" [*Sic! Sic! Sic!*][2]

When I am feeling low, I turn for spiritual sustenance to the *Bull Cook Book* by George Leonard Herter [Waseca, Minn: Herters Inc., 1963]. It is not the quality of the recipes that I admire (although

"Spinach Mother of Christ" has a certain *je ne sais quoi*), nor do I necessarily agree with the aesthetic, social, or political dicta with which the text is interlarded. It is the *tone* of the work, the vivid impression it conveys of an author brimming over with strongly held opinions, some of them surprising, and with facts not generally known (e.g., "Mary, the mother of Jesus, was very fond of spinach. The way she liked it best was ..."), all of it uttered with sublime assurance and self-confidence. Those wimps, Uncertainty and Doubt, cannot long survive in the company of Mr. Herter.

Robert Ferguson, a Texas polygraph examiner of long experience, was made by the same firm that gave us Herter, and since this is Ferguson's fifth book on polygraphy, I have had the pleasure of his company before. Chris Gugas, Sr., another elder statesman of the lie detector industry, has also published previously,[3] but the style of the current volume is so clearly Ferguson's that one suspects Gugas's role was mainly inspirational. His name also lends dignity to the book jacket by virtue of the Ph.D. degree (conferred by the University of Beverly Hills, an institution unknown to the twelfth edition of *American Colleges and Universities*). For a man like Ferguson, with resonant opinions and a taste for zesty language, what a joy it must be to see his views set up in print. Let us relish a few examples:

> Unfortunately, the minute a small handful of psychologists—one or two pseudo-knowledgeable and one or two completely ignorant of what they were even trying to do—got into the picture, two expressions, "false positive" and "false negative" came to light. It appears that some people turn out to be weird ducks. Sadly, when that type of inquirer doesn't understand something, he is usually prone to attach strange names to it under the guise of professionalism or scientific exploration on both sides of the same coin. By confusing other people more so than himself, he feels he can still call himself an "expert." Those two phrases appeared in a tumor in the brain. Before then, they had never existed in polygraph language. In all sincerity, however, foul ball psychologists are few and far between. (pp. 25–26)

That final ameliorative sentence is characteristic; one can picture Ferguson getting madder and louder, pounding the typewriter, until it occurs to him that some readers might think he is going too far. Through clenched teeth, he squeezes out that brief tergiversation, which he obviously does not believe himself. In a fulmination covering three pages, Ferguson explains who is to blame for the "cost burden" to private employers of employee theft and turnover:

Much of it is founded in that large, irresponsible segment of youthful society [for whom] literally anything, even registering for the draft, that gets in the way of their self expression ... or purports to obstruct their immature understanding of that mythical misnomer termed "personal privacy" (i.e., employer rules and regulations) turns them off.... Then there are the professional job jumpers who keep bounding only to figure out another phony on-the-job workman's compensation claim.... Next come the chronic alcoholics who are constantly sneaking off some place for a little nip.... Let's move on to the pot heads, needle freaks, and pill poppers.... Not to be ignored is the expert "impersonator," smooth and personable.... Then comes the amateur thief ... and finally the professional thief.... Private industry is loaded with the dishonest. Wholesale and retail outlets are prime targets ... many a man's word is no longer his bond as it pridefully was ... [due to] the prevailing element of self-centered irresponsibility [and] monumental employee dishonesty. (pp. 18–20)

All of this creates a very bleak picture indeed. If this is what things are like in Ferguson's neighborhood, I don't think I want to move to the Sun Belt after all. Nor am I reassured when he tacks on another of those dubious disclaimers: "Fortunately, most people are good and most people are basically honest. Unfortunately, some are not ..." (p. 21).

These samples are from a chapter splendidly titled "The Real Truth of the Matter." Some equally good points are made in Chapter 8, "Eliminating Adverse Potentials," for example:

Any form of restrictive polygraph legislation existing or possibly pending suggests nothing more than political chicanery, most likely unconstitutional from numerous viewpoints.

In Chapter 6, "Preemployment Psychological Concomitants," we learn that "emotions make us happy and sad," and we are enjoined to "remember, all behavior has a cause." We also learn in Chapter 6 that the real "*lie detector*" is not the polygraph instrument nor even the polygraph examiner; the real lie detector is in fact the "testee himself."

Ferguson's explanations of assertions like this last one are not always clear; for example:

One individual life in itself, and of itself, is important only to the beholder thereof. If we eliminate the corollary of outside emotional factors temporarily attached to one individual's life span, we further discover in a rather cold-blooded context that another person's life is absolutely of but momentary and temporary importance to you! Shocking but true. And that is why man is his own lie detector, in every unequivocal sense of the word. (p. 76)

To be fair, there is another way of explaining this claim (that the "testee himself" is the lie detector), which I am sure Ferguson would have offered if he had not been distracted by these disturbing questions of mortality. He and Gugas advocate a polygraph test format called the Relevant–Irrelevant technique, in which innocuous questions— "Is today Sunday?"—are intermixed with relevant questions— "Have you ever stolen money from a previous employer?" "Have you ever been arrested?" "Have you suffered a back injury?"—while the "testee" is attached to a polygraph. If the subject answers any of these questions deceptively, it is assumed that he or she will know it and will reveal it through the response recorded on the polygram. Thus does the subject become his or her own lie detector, and this "fact" is "completely supported by many, many endless forms of scientific proofs" (p. 75).

When one or more of those questions makes his palms sweat or his heart beat a little faster, the machine is turned off, the examiner probes:

> I'm not trying to embarrass you but something about the arrest and injury questions has caused chart responses here. Do you want to tell me about them? What was going through your mind when you answered the arrest question? (p. 114)

Job applicants often make damaging admissions at this point, trying to account for these alleged signs of deception, accepting the examiner's invitation to be open and forthcoming. These admissions, especially as paraphrased in the examiner's report, may sound quite sensational to prospective employers, who seldom stop to wonder what they themselves might admit under similar pressures, and they are grateful to the polygrapher for warning them away from that applicant.

Is this sort of ordeal an invasion of privacy?

> This is utter nonsense. First and foremost, there has never been a constitutional or legal definition of privacy. The old hue and cry of "invasion of privacy" perpetuated upon [sic] and truly worn out by the more left wing, radical critics of polygraphy, has no basis in fact or at law. (p. 151) [Q.E.D]

Is Ferguson's kind of self-assurance innate, or can it be acquired? Surely years of sitting, Godlike, in judgment upon frightened and often undereducated people must have some cumulative effect? You have Charlie alone in your office, eyeing that scary-looking machine on your desk. He has already signed a release form that seems to waive all of his rights and permits you not only to ask him whatever you want to ask about his history, his sins, and what he is thinking but also to draw conclusions about his character and honesty and report them to his boss.

> If the one-in-five-thousand smart aleck subject says, "I'm not going to sign that," simply explain, "No signee—no testee." Send him on his way and forget the nut. (p. 38)

Ferguson sees an analogy to the relationship of a doctor with a patient (p. 21); the polygrapher apparently is a kind of surgeon of the soul. If it is power that rings your chimes, the lie detector business is a treat.

Ferguson's self-confidence is not without limits, however. After he has decided that Charlie is not deserving of employment in Fort Worth, Ferguson is prudent enough to relay his conclusion to the prospective employer by telephone and to keep no written record. "The modern examiner will discourage written reports in routine preemployment and periodic testing as much as possible." (p. 139). Come to think of it, I guess I like "Bull Cook" Herter's brand of self-assurance better, on the whole, and also his profession.

The Format of the Screening Test

In the screening situation the respondent is not suspected of some specific criminal act and therefore it is not possible to design specific relevant or control questions. The so-called Positive Control Test can employ both general questions ("Have you ever taken money from your employer without permission?") and specific questions ("Did you take the missing camera?") and is sometimes used for screening purposes. But by far the most widely used screening format is the one now to be discussed. The questions used in these screening tests will vary with the examiner and, especially, with the particular application. A typical preemployment screening test similar to those used by the FBI will contain two or three irrelevant questions and numerous relevant questions touching on various areas of interest to the prospective employer. Because of the lack of control questions of either the known-lie or the known-truth variety, most examiners are inclined to identify these screening tests with the old relevant/irrelevant format. Because of the extreme implausibility of the assumptions underlying the R/I test, this screening procedure is considered to be less accurate than the CQT and suitable primarily as a device for inducing admissions.

The classical R/I test as used by Larson and Keeler in criminal investigations used relevant questions that all focused on the specific issue or crime of which the respondent was suspected. Therefore, innocent as well as guilty suspects were likely to respond more strongly to relevant than to irrelevant questions and thus fail the test. But the numerous relevant questions of a typical screening test deal with many separate issues, and it would seem unlikely that any one individual would feel compelled to answer all of them deceptively. Therefore, it might be reasonable to use some of these relevant questions as controls for other relevant questions. That is, if a given subject responds more strongly to questions about drugs than he or she does to any of the other questions, then we might be inclined to agree with the examiner's conclusion that, "I think you are having a problem about the drug questions, George. Can you tell me what might be causing you to react this way?" If all of the relevant questions referred in some way to drug use, then stronger reactions to these than to the irrelevant questions would be intractably ambiguous. But here the other relevant questions also refer to sensitive areas, have the same accusatory implications, yet our subject reacts differentially to the one involving the area of illegal drugs. The proper comparison, and the one that no doubt is actually used by polygraphers engaged in such screening, is between responses to the different relevant questions, and this procedure might be more accurately described as a Relevant Control Test (RCT). However, we are accumulating too many confusing acronyms so I shall speak simply of the screening test format.

In the screening situation, the examiner will go over the questions with the subject before attaching the polygraph. The subject is made to understand that he is supposed to be able to answer each of the relevant questions truthfully in the negative. If he thinks he might have difficulty with any of the questions, he is encouraged to explain why so that the wording can be changed to exclude "what you have already told me." A careful examiner will use at least three question lists covering the same topics but in different words and with a different ordering. As we shall see, the main purpose of this screening test is to elicit admissions, and when still permitted in the private sector, it was relatively infrequently that employers received adverse reports based entirely on the polygraph record. But our present concern is with the validity of the screening procedure as a test because, as we shall see, the FBI, the military, and probably all the federal intelligence agencies use it in exactly that way. If the respondent persistently shows greater polygraphic reaction in one or two content areas than he does in the other areas, it is concluded that he has been deceptive or that some unexplained problem disturbs him about those one or two subjects.

Assumptions of the Polygraph Screening Test

ASSUMPTION 1. *To a truthful subject who genuinely has no "problem" in relation to any of the areas covered, the various relevant questions will all seem about equally threatening or disturbing.*

If someone reacts selectively to questions dealing with his or her sexual behavior, one could not reasonably infer that this was evidence of deception. Some content areas, and some forms of question, are intrinsically more disturbing than others. Ask a psychiatrist to formulate two or three accusatory questions that will have exactly equal stimulus value (that will produce just the same degree of response) for some patient he knows intimately and he may protest that you are asking too much of his knowledge and skills. Ask him to generate a list of 10 or 15 questions that will be similarly homogeneous in their impact on people in general and he may well throw up his hands. Formulating a suitable screening question list is not a trivial task.

ASSUMPTION 2. *If a subject has offered admissions in some area, then changing those questions to the form, "Apart from what you have told me ...," will remove their heightened stimulus value for that subject.*

This seems to me to be a risky assumption. If, in his eagerness to get a clean bill of health from the polygraph, the subject admits to, say, a prior arrest that he is embarrassed about and did not mention on his application form, then it would not be surprising if this same subject continued to respond to the modified question even though his denial now is truthful.

The question still touches the same painful nerve and continues to produce a twinge.

ASSUMPTION 3. *Given a properly balanced question set, as assumed in #1, then persisting strong polygraphic responses to one or two content areas, as compared to that subject's response to other areas, indicates that the subject is being deceptive in the reactive areas.*

I consider this assumption plausible but overstated. One can think of various alternative explanations for such selective responding. A young woman might think that the only misdeed listed of which she could possibly be suspected would be some minor theft; therefore she might respond selectively to the theft questions without having actually been guilty of stealing. A young man might be sensitized to the drug questions because he was one of the few among his friends who had *not* experimented with street drugs rather than the other way around. Another youngster might regard the questions about being sent by "some outside group" to join the FBI, or being wanted by the police, as being so implausible in relation to himself that he reacts selectively for that reason. Or the examiner, quite inadvertently, may emphasize one question by his tone or timing and thus ensure that the subject will show a heightened reaction to that topic henceforth. Persisting, selective responding in some content area raises a question for which deception may be a plausible explanation, but it will seldom be the *only* plausible explanation.

I should mention here that at least some, perhaps most, federal polygraphers customarily make an egregious mistake that wholly invalidates Assumption 3. When a subject reacts more to one of the questions than to the others—in my experience, the questions about illegal drug use most commonly produce such reactions on federal tests—the examiner typically comments: "I'm getting a reaction to this question about marijuana. Is there something you ought to tell me about that?" However this conversation may conclude, it is to be expected that, on the next run through the question list, that subject will respond even more strongly to the marijuana question, which has now been singled out as a real trouble spot. As we shall see, one FBI candidate, Elizabeth M., was told before her screening test that "one-third of FBI candidates fail on the drug questions." It is difficult to think of any wrongdoing, no matter how unlikely, that could not be turned into an excitatory stimulus by the news that one had reacted to it in a way that the examiner considers to be suspicious: "Are you a member of the Mafia?" "Are you employed by the KGB?"

Sometimes a thought experiment can be helpful. Suppose we were to administer polygraph screening tests to 20 consecutive applicants for jobs as FBI agents. They would all be well-educated and well-socialized young people, with honorable records and excellent letters of recommendation. Before each candidate arrives for the test, we randomly select one of the topics normally used in the screening process, questions about the use of illegal drugs, lying on the application form, sexual misconduct, theft, and so on. During the pretest interview, we shall seem to be especially concerned about our selected issue; we appear doubtful when the subject denies that behavior; our formerly friendly manner changes to one of apparent suspicion. During the first run through the list of questions with the polygraph in operation, we pause an unusually long time after the subject answers our target question. Then we stop the polygraph, relax the pressure in the arm cuff, and sit down in front of the subject. "I'm getting reactions to the [selected issue] question, John. Is there anything you want to tell me about that?" We appear not to believe his denials but finally we continue with a second run through the same questions. During the next rest pause, we tell the subject: "John, we really have a problem with the [issue] questions here. Anyone looking at these charts would conclude that you are not telling me the truth about this. I strongly advise you to come clean now, tell me whatever it is that's causing these reactions, and then it is at least possible that the last chart will be better." And so on. *Question*: Is there any doubt that we could cause most or all of these young applicants to respond significantly to our target questions after this type of preconditioning? *Answer*: No, there is no doubt at all.

The Validity of the Polygraph Screening Test

Remarkably, in spite of their widespread and growing use, there has been *no* published research on the accuracy of screening tests of this type. The fact that American business eagerly invested millions during the 1970s and 1980s to pay for polygraph screening has no evidentiary value. Employers were primarily interested in the admissions that are elicited in the polygraph situation and in the assumed deterrent value of periodic testing. Neither of these consequences requires that the test be valid but only that employees *think* it is valid. Polygraph examiners, even those who have administered thousands of such screening tests, have no way of knowing how accurate they are since, in the vast majority of cases, they will never have independent evidence as to which subjects were lying and which truthful.

In 1979, the Oversight Subcommittee of the Select Committee on Intelligence of the U.S. House of Representatives learned that polygraph testing is a central component of the preemployment screening given to job applicants in most federal police and security agencies. At least 75% of those refused clearances by the CIA or National Security Administration were turned down because of lie detector test results. (In CIA slang, to be given a lie test is to be "fluttered.") Concerned by the total lack of evidence for the validity of these procedures, the subcommittee urged the director of the CIA to institute research on "the accuracy of the polygraph in the pre-employment setting and to establish some level of confidence in the use of that technique."[4] No credible research on this important topic, however, has as yet been published.

Verdict

No one knows whether the screening test has some, slight, or no validity at all. When asked by prospective employer-clients whether the test was accurate, the polygrapher's only honest answer would have to be, "Nobody knows." (Mr. Ferguson's answer, on the other hand, would be that the validity of this technique is "completely supported by many, many endless forms of scientific proofs"—proofs that, sadly, exist only in Mr. Ferguson's imagination.) Considered strictly as a test, the assumptions of the screening procedure seem actually to be less implausible than those of either the R/I or the CQT methods; but while not blatantly implausible, the assumptions are wholly untested by experiment and are certainly untrue in some cases. Perhaps most importantly, however, as our thought experiment discloses, any possible validity of the screening test is at once compromised unless the examiners are known to be scrupulous and skillful in their test administration. The examples given in Chapter 15 of flagrant examiner misconduct on the part of federal examiners, whom one would expect to be among the best in the business, strongly suggest that polygraph screening, as actually practiced, is "20th century witchcraft" indeed, just as the late Sam Ervin so presciently suspected.

Chapter 11

VOICE STRESS ANALYSIS

Speech is the mirror of the Soul: as a man speaks, so is he.

—PUBLILIUS SYRUS, Maxim 1073

Chicago Brothel's Lie Detector Fails to Uncover Undercover Cop. A lie-testing apparatus called a "psychological stress evaluator" ... was among items seized in a vice raid by police at a place called the Quest in suburban Des Plaines. Earlier, an undercover sleuth ... was required to submit to such a test while posing as a seeker of a prostitute's services. (Later) the officer was informed ... that he had passed the test and was eligible to avail himself of the Quest's ... playmate. As she disrobed, the (police) raiders struck ...

—*Chicago Sun-Times*, August 17, 1979

During the 1960s, the U.S. Defense Department spent hundreds of thousands of dollars in an eager search for feasible methods of covert lie detection, techniques by which bodily reactions related to stress might be measured without the subject's knowledge. Explaining its interest to holders of research contracts, the department alluded vaguely to cloak-and-dagger situations in which an American intelligence officer might be receiving a report from one of his agents who, in turn, might secretly be working for the Other Side. One is entitled to suspect, however, that another source of concern was the adverse verdict of the Moss Committee, a subcommittee of the House Governmental Operations Committee, which had criticized in 1964[1] the unbridled use of polygraphic interrogation by federal agencies. If a covert technology could be developed, then fewer aggrieved victims of the lie test would complain to their congressmen. Considerable interest focused for a while on the "fidgetometer," or "wiggle seat," a chair fitted either with pneumatic cushions or strain gauges in the legs and arms to provide secret recordings of the subject's movements during questioning. One version of the wiggle seat could also provide rough measurement of body temperature. Another line of research concerned changes in pupil size as recorded by hidden movie cameras.

Perhaps the most remarkable development was based on the extraordinarily sensitive infrared (IR) detectors that had been created for use as the sense organ of the heat-seeking missile. I was given one of these devices to try out in my laboratory. It could be aimed through a hole in the wall at the upper lip of a subject 20 feet away and produce a clear record of his breathing (the upper lip alternately cools and warms slightly as the subject breathes in and out through his nose) as well as baseline changes due to flushing or blanching of the skin. No one worried much about what we might do with these measurements once we had them, whether they would actually give valid evidence about the subject's veracity. Attention focused instead on the technological problems, which were more glamorous and also more soluble. In the case of the IR detector, the great problem was how to keep the thing trained on the upper lip of an unsuspecting subject who doesn't sit perfectly still. The answer, which I saw only on the drawing board, would be highly gratifying to any space-age engineer. Behind that hole in the wall, mounted on a fancy gamboled tripod, would be a four-barreled device consisting of two IR detectors, one ordinary telescopic gun sight, and a laser. With the gun sight, the operator aims the main IR detector on the subject's upper lip and pulls the trigger. This causes the laser to project a microsecond beam at a spot on the subject's cheek, imperceptibly warming that spot by a degree or two. The second IR detector is narrowly trained on that warm spot and follows it, driving a marvelous system of hydraulic valves and pistons that moves the detector as necessary to keep the warm spot in view of IR #2 which, in turn, keeps IR #1 trained on the upper lip. Periodically, as the warm spot cooled, the laser would automatically be triggered to warm it up again. What a splendid conception! Who could quarrel with the expenditure of a few hundred thousand of our tax dollars to perfect such an elegant piece of hardware before the Russians got onto it?

A Tremor in the Voice: The PSE

It remained for two Army officers, however, to perceive the obvious approach to covert lie detection: *get it from the voice!* Who can deny that the voice changes under stress? We have all heard it in our own voices when we are frightened or angry or sad, and we have heard it in the voices of our friends. We don't need fidgetometers or cameras or holes in the wall, just a hidden microphone and a recorder, with which one could detect stress even over the telephone, the radio, or television. By working with the recorded speaking voice, one could even apply stress (lie) detection to the testimony of the dead as was demonstrated by one enthusiast who analyzed sound recordings obtained from Lee Harvey Oswald during the short time between his arrest and murder, "proving" that Oswald was innocent of killing President Kennedy. Lt. Col. Charles McQuiston, an Army polygrapher, and Lt. Col. Allen Bell, an intelligence officer, retired from the service in 1970 and established Dektor Counterintelligence & Security, Inc., a Virginia corporation devoted to the manufacture of the Psychological Stress Evaluator, popularly known as the PSE, and to the training of PSE examiners in a five-day course that is included in the price of the instrument.

The idea that the speaking voice changes under stress is reasonable enough but several practical questions remain. First of all, how does the voice change? Speech sounds are exceedingly complex. Only after many years of effort were scientists at Bell Telephone's Haskin's Laboratory successful in programming a computer to produce lifelike spoken words and phrases. Many aspects of the voice might change with stress; which ones do? Second, do different types of stress produce the same or different changes in a given speaker's voice? And third, does the same type of stress yield the same types of change in different speakers? We know that some people's hearts speed up in anticipation of a painful stimulus, whereas others' hearts slow down; do voice changes show similar idiosyncrasies? Finally, of course, if we *could* measure stress from the voice, and our measurements meant the same thing for different people, how could we use these measurements for lie detection? Just like the polygraph, all that any voice stress analyzer could hope to do would be to show that the subject was more (or less) aroused or "stressed" when he replied to this question than he was when he replied to that one. Any voice stress device could be merely another polygraph variable or channel, one that could be obtained covertly or at a distance, but its use in the detection of deception would be subject to precisely the same problems and limitations that affect the variables measured by the polygraph.

Dektor said that its PSE measured slow (10 per second) variations in the frequency or pitch of the speaking voice. It had been established by a British physiologist named Lippold that the voluntary muscles in the arm generate a tiny physiological tremor or microvibration at about 10 per second when the muscle is at rest and the subject relaxed.[2] When the subject is stimulated or aroused, the microvibration tends to fade or disappear. The theory behind the PSE and later devices of its type is that the muscles in the throat and larynx may also show this microvibration and may communicate it to the voice by a process of frequency modulation so that, in a relaxed speaker, the voice would show a maximum of this 10-per-second warble, while a stressed speaker would show a minimum. Consistent with the priorities that seem characteristic of the lie detector business, Dektor had been selling PSE machines by the hundreds for six years before the first bit of evidence appeared for this interesting theory. Inbar and Eden, two electrical engineers in Israel, found 10- to 20-per-second frequency changes in the voices of five subjects and were also able to record simultaneous electrical indications of physiological tremor from the throat muscles of these subjects.[3] The correlation between these two, however, needed to infer that the voice vibrations are caused by the muscle tremor was very marginal. Whether the instrument manufactured by Dektor is capable of detecting these voice changes has yet to be demonstrated by independent research.

Does the PSE Measure Stress?

If the PSE is to be useful in lie detection, a minimum requirement would be that the instrument can detect simple differences in levels of stress. In 1974, Brenner analyzed recordings from subjects who were required to give dramatic readings before small audiences and found a slight (.32) correlation between their self-reports of stage fright and the PSE stress levels. There was also an increase in PSE score with the size of the audience.[4] Smith found "high stress" in the voices of 6 of 13 persons calling in to a radio talk show, in 7 of 8 phobic patients asked to count aloud into a microphone, and in 10 of 25 "normals" and nonphobics given that same laboratory task.[5] Using mental arithmetic tasks of varying difficulty, Brenner, Branscomb, and Schwartz did a PSE analysis of the digits spoken as answers to the problems and found that the harder tasks showed higher PSE scores than the easier ones.[6] (They also found that the spoken digits 5 and 9 consistently showed higher PSE stress scores than the digits 2 and 8. "Stress" scores based on different words are apparently not comparable to one another.) On the negative side, Lynch and Henry asked 43 subjects to read into a microphone a mixed list of neutral and "taboo" (naughty) words, on the assumption that when, in the sober confines of the laboratory, one says "s--t," there should be more stress in the voice when the middle letters are "hi" than when they are "oo." Two PSE-trained analysts and 10 amateur chart readers were all equally unable to differentiate the taboo from the neutral words on the basis of their PSE readings.[7] As part of a major investigation conducted for the Israeli police, Nachshon asked 20 subjects to view a series of slides, half of them pastoral landscapes and the other half color photographs of mutilated corpses (the experiment was referred to as the Horror Picture test). The subjects contemplated each slide for 5 seconds and then, on cue, spoke the words, "Yes, I like this picture," into the microphone. *Question*: Could PSE analysis of these utterances distinguish those spoken while subjects were viewing the attractive landscapes from those uttered while they contemplated corpses? *Answer*: No. Three PSE chart analysts independently studied the resulting records, and part of the problem can be seen from examining the interjudge agreement. In guessing from the PSE charts which utterances had accompanied the horror pictures, no two judges agreed more than 65% of the time and all three agreed only 40% of the time. In this experiment, sorting the charts at random should have produced 50% correct classification, but the three judges managed only 50%, 40%, and 35% respectively.[8]

This same lack of reliability—interscorer agreement—has been observed in several other studies summarized in Brenner's testimony before a Senate subcommittee in 1978.[9] If a test is not reliable, it cannot be valid.

Does the PSE Detect Lying?

Given that the PSE cannot seem to dependably distinguish words or phrases uttered under conditions of high versus low stress, it may seem supererogatory to look into its functioning as a lie detector. But there have been not only claims of high validity, but also reports of actual "studies" in which the PSE is said to have produced outstanding results. A. E. Dahm, a Dektor employee, mailed questionnaires to 423 users of the PSE and reported to a congressional committee in 1974 that the 46 who had replied seemed to think highly of the instrument and its capabilities. A Maryland policeman named Kradz, later Dektor's chief instructor, reported (to Dektor) in 1972 that he had tested 42 criminal suspects with both the PSE and the polygraph and obtained 100% agreement with another examiner on PSE scoring and also 100% validity as measured by agreement with the results of independent verification of the suspect's actual guilt or innocence. Finally, there is John W. Heisse, Jr., a physician who was president of the International Society of Stress Analysts (ISSA; an organization of PSE examiners) and who ran a lie detection service in Burlington, Vermont. Dr. Heisse submitted to a Senate committee in 1978 the report of an unpublished study that claims to have achieved 96.12% accuracy by blind readings of PSE charts obtained from 52 criminal suspects about whom ground truth had been established by a variety of means.[10]

The Dahm report does not deserve serious attention, and one can only wonder at the marvels achieved by Lt. Kradz. The Heisse study is of a type that we have seen before (the four lie test studies by the Reid group discussed in Chapter 8), and it is worth thinking about how it was designed—perhaps as follows: The president of the ISSA writes to 11 of his fellow PSE examiners, requesting that they send him charts obtained from verified-innocent or verified-guilty suspects. He explains that these are to be blindly scored by five other PSE experts in order to assess the reliability and validity of the technique on which they mutually depend for their livelihood. As is also true for the polygraph, specific PSE patterns are said to indicate high or, conversely, low stress. Lie test charts showing these established high-stress indications in the answers to relevant questions, and low-stress indications associated with control or guilt complex questions, can be referred to as classic deceptive charts; any graduate of the five-day Dektor training should be able to identify such charts as indicating deception. Similarly, charts showing low-stress indicators on the relevant questions, or higher stress indications on the control questions, can be referred to as classic truthful charts, which any PSE practitioner should classify as truthful. In any proper study of the reliability and validity of the PSE as a lie detection tool, the question we must ask can be stated thus: What proportion of the PSE lie test charts obtained from a representative or random sample of criminal suspects, verified as to guilt or innocence, will be of this classic type, which all PSE examiners should score the same and should score correctly? In respect to the Heisse study, the question we must ask is, "Is it probable that charts obtained as his were constituted a random sample of criminal suspects?" Put another way, how likely is it that any examiner solicited in this manner would contribute for study a chart from a guilty suspect that did *not* conform to the classic deceptive standard or charts from innocent suspects that did *not* fit the classic truthful model?

None of the affirmative evidence suggesting that the PSE is useful in lie detection has been published in an edited scientific journal. Quite apart from its provenance of vested interest, none of it meets minimum scientific standards. Let us now turn to some research that does meet such standards. In 1973, Professor J. Kubis of Fordham University conducted a comparison study of conventional polygraphy with results based on the PSE and also on a voice stress analyzer, another device intended to assess stress from voice frequency changes.[11] This was a laboratory experiment involving a mock crime situation in which one subject acted out the role of "thief" while another served as "lookout," or accomplice. A third subject in each triad was "innocent," that is, did not participate in the enacted theft. Since the requirement here was to classify subjects as guilty, accomplice, or innocent, the chance rate of success would have been 33%. The better of the two polygraphers achieved 61% correct classification and, as is usual, did better at detecting the guilty (85%) than the innocent (54%). When allowed to compare the charts of the three members of a triad (so that he did not have to ask whether an individual chart appeared deceptive or truthful, but rather which of three given charts looked *most* deceptive or *most* truthful), the polygrapher, of course, did better, with an average of 76% success on 45 such triads. The immediate global impression of the examiner who administered the tests was correct 65% of the time and the clinical impressions of judges who merely listened to the tape recordings of the interviews were correct 55% of the time overall. The PSE, in contrast, achieved only 32% accuracy in classifying 85 subjects (some recordings were not good enough for PSE analysis) and only 38% when the voice tapes were considered in triads. The Voice Stress Analyzer similarly managed only 35% hits in the triad analysis. Both of the voice analyzers, therefore, performed about as well as one might do by using random numbers.

Nachshon, in 1977, examined the ability of the PSE to identify the card chosen by a subject in the typical card test.[12] (Subject picks one of six cards, each bearing a different number, answers "No" to questions of the form, "Did you pick card Number X?") Using 20 college students as subjects, the average interjudge agreement on the number chosen, based on the PSE

charts, was again very low, not over 30%. Similar findings were obtained
on card tests administered to 19 criminal suspects as part of regular polygraph examinations. Horvath recently used the card design with 60 volunteer subjects, comparing the ability of trained examiners to identify the
chosen card from PSE charts versus electrodermal response (EDR) records
obtained with a standard polygraph.[13] Since there were five cards to
choose from, chance "detection" rate was 20%. Using the EDR, two evaluators averaged 56% correct calls, but got only 23% using the PSE, no
better than chance. Brenner, Branscomb, and Schwartz tried out the PSE in
a guilty knowledge test design similar to one in which I had previously
been able to "detect" 20 out of 20 subjects on the basis of EDR records.[14]
The PSE, however, again did no better than chance.[15]

How does the PSE perform in the serious business of real-life interrogation of criminal suspects? Barland, in 1975, made simultaneous PSE
and polygraph recordings in CQT lie test examinations of 66 criminal
suspects.[16] The PSE decision as to truthful or deceptive agreed with the
polygraph outcome 53% of the time and with independent judicial outcome 47% of the time, where 50% was chance expectancy in both cases.
(An earlier report by Barland in 1973, based on a subset of these cases, had
suggested positive findings, but this conclusion turned out to have resulted from a statistical error.) Tobin, using an Israeli modification of the
PSE by Inbar and Eden, examined 32 criminal suspects with only chance
results.[17]

Nachshon made audio recordings of 41 polygraph examinations being routinely conducted by the Israeli police on criminal suspects. Three
judges independently evaluated the PSE charts. The average agreement in
scoring between pairs of judges was 51% (i.e., chance), and the average
agreement between the PSE scoring and the conventional polygrapher's
evaluation was also 51%, or chance. The one apparently nonchance finding
in this study was that, while the polygrapher classified about two-thirds of
the suspects as truthful, the PSE evaluators classified two-thirds of them
as deceptive. Nachshon concludes that

> the PSE is neither reliable nor valid for detecting emotional arousal
> [and] the findings of the field study further imply that PSE employ-
> ment in real life situations is not only useless, but also dangerous
> [because] the judges, when making an incorrect evaluation, tended to
> incriminate the innocent rather than to clear the criminal![18]

In 1980, a complete Psychological Stress Evaluator cost about $4,400
f.o.b. from Dektor Counterintelligence, Inc., then located in Savannah,

Georgia. As we have exhaustively determined, using the PSE to differentiate levels of stress or to detect deception produces results that typically conform to what might more easily be achieved by flipping a coin. If one were to use a "silver" dollar coin, the net saving would be $4,399. It must be remembered that *hundreds* of these machines have been purchased and are in daily operation in the United States, separating the "liars" from the "truthful," deciding who will be hired and who fired. And most of these PSE examinations are being paid for by American businessmen, apparently the same hardheaded, clear-eyed managers who helped Richard H. Bennett, Jr., to become a millionaire.

Bennett was the president of Hagoth Corporation, manufacturer of a $1,500 device widely advertised in airline magazines with which, it was claimed, the businessman can determine truth from falsehood in office or telephone negotiations. The great thing about the Hagoth machine, apart from the sensational advertising campaign, is that it leaves no written record; the only output is an evanescent display of winking lights, some green (for "low stress") and some red ("high stress"). Therefore, when the businessman discovers that he has made an enemy through unwarranted suspicion or been swindled through Hagoth-induced credulity, he has nothing but a hazy memory of reds and greens to blame it on. Asked for evidence that the Hagoth works, Mr. Bennett breezily replied, "I'll show you my bank balance."

Recent Developments

Since the above appraisal was published in 1980, scientific interest in voice stress analysis has waned but not that of the credulous public. The Hagoth has been replaced by the Truth Phone, a $4,000 gadget that provides "digital lie detection for analyzing the voices of those you speak with on the phone. Digital read-out alerts you of possible deception."[19] The PSE has been replaced by the Computer Voice Stress Analyzer (CVSA) produced by the NITV company of West Palm Beach, Florida, and is sold to law enforcement agencies for $8,400 together with a six-day certified examiner's course that costs $975 per student. The CVSA is a modified "notebook" computer with a microphone attachment, and the microtremor content of the subject's spoken answers is displayed directly on the computer screen. When asked about validity studies of the CVSA, Dr. Charles Humble, executive director of NITV, replied:

> Insofar as "research" that has been conducted on the CVSA, we de-
> cided very early on not to follow in the same footsteps as the poly-
> graph only to find ourselves in the same position the polygraph finds
> itself in today. The CVSA is sold as an investigative tool and nothing
> more. Our students are taught that it is not to be used in court, to obtain
> warrants, or to in any way persecute anyone. The system was designed
> to identify innocent individuals and help remove them as suspects. All
> of our instructors teach that if a person is not able to "pass" the exam, it
> simply means that they are not eliminated as suspects—period.[20]

I interpret this statement to mean that CVSA users are encouraged to use
the device as an interrogation tool for eliciting confessions (this is the real
utility of "lie detectors" although it has its perils; see Chapter 16) rather
than as a test per se. However, the CVSA literature alleges that the device
has been shown to have "Accuracy (98%)," which sounds like the claims
made by polygraphers. Nevertheless, the CVSA brochure includes numerous letters from satisfied customers in police departments around the
United States describing cases solved by confessions obtained from suspects during CVSA interrogations. One innovation by NITV has been to
encourage police investigators, rather than polygraphers, to take their six-
day course and then to use the instrument themselves. This is a wise
business practice because attempting to persuade polygraphers to abandon methods they already believe to be nearly infallible would be unlikely
to succeed. It also means that the investigators who do the initial interviews with potential suspects can employ the "bloodless third degree" of
instrumental lie detection *ab initio*, when it is most likely to be useful.

Because of the CVSA's growing popularity, the Department of Defense's Polygraph Institute (DoDPI) at Ft. McClellan, Alabama, purchased
a CVSA and sent two of its researchers to the CVSA school for training
by the manufacturer. DoDPI then conducted its own in-house research
leading to a position statement published in 1996 that concludes, in part:
"We found no credible evidence in information furnished by the manufacturer, the scientific literature, or in our own research, that voice stress
analysis is an effective investigative tool for determining deception."[21] The
polygraph unit of the Los Angeles County Sheriff's Department reports a
similar experience.[22]

Verdict

There is no scientifically credible evidence that the PSE, the CVSA, the
Mark 1000 VSA, the Hagoth, the Truth Phone, or any other currently

available device can reliably measure differences in "stress" as reflected in the human voice. There is considerable evidence that these devices, used in connection with standard lie detection test interrogations, discriminate the deceptive from the truthful at about chance levels of accuracy; that is, the voice stress "lie test" has roughly zero validity. One business enterprise to learn this, to their cost, was the high-tech house of prostitution mentioned in an epigraph to this chapter. According to a 1979 story in the *Chicago Sun-Times*, an undercover agent for the Cook County sheriff's vice squad was required to submit to a PSE lie test when he visited the house posing as a client. The test questions had to do with whether he was connected in any way with the police. To his surprise, the agent passed the test and was granted client privileges. The PSE was confiscated in the ensuing sheriff's raid but the news report does not reveal what finally became of it. Let us hope the Cook County sheriff is not using it to interrogate criminal suspects.

Although, on second thought, if I were running a homicide or major crimes unit of a large metropolitan police department, I would be tempted to buy a CVSA and have several of my detectives trained in its use. I would encourage them to use it in questioning suspects and see that accurate, standardized records were kept of the outcome of each interrogation. I would make it clear that no suspect was to be considered permanently cleared merely because he "passed" the CVSA test and that no confession was to be accepted as truthful unless it could be independently verified. In short, I would make sure that the technique was employed as an interrogation tool and not as an actual test of truth. For reasons more fully developed in Chapter 16, I would expect this experiment to show a useful payoff in more cases cleared more expeditiously.

Chapter **12**

SURVEYS OF SCIENTIFIC OPINION OF THE "LIE DETECTOR"

There is another way to truth: by the minute examination of the facts. That is the way of the scientist: a hard and noble and thankless way.

—JOHN MASEFIELD

As we have already seen, fully adequate research-based estimates of polygraph accuracy in real-life applications are not yet available. There is strong reason to distrust laboratory studies of the CQT in which volunteers are required to commit mock crimes and lie about them on a subsequent polygraph test. Moreover, field studies in which the only criterion of ground truth is confessions induced by interrogation after failed CQTs *must* overestimate the true accuracy of the technique for the reasons reviewed at length in Chapter 5. Because these studies agree in showing that the CQT does only slightly better than chance in affirming the truthfulness of innocent suspects, we can reasonably assert that failed CQTs have negligible probative value. But these same studies suggest that the CQT has an accuracy of over 85% in detecting lying. If we could rely on this finding, then there might be justification for permitting juries to learn that a criminal defendant had passed a CQT, and also for the CIA and the FBI to continue to require their agents to pass polygraph tests.

In addition, while considerably smaller now than it was before the federal Employee Polygraph Protection Act was passed in 1988, the lie detector industry remains firmly entrenched in local and state police departments around the United States, in the military, and in the federal bureaucracy. Although its supporters within the scientific community are few, some are quite vocal and willing to vigorously dispute most of the arguments made in this book. They will insist that the laboratory studies by Raskin and colleagues, which purport to show extraordinary validity for the polygraph, can be extrapolated to real-life applications. They will deny that the assumptions of the various forms of lie detector testing are as wildly implausible as I have characterized them to be. By marrying the myth of the lie detector with the mystique of the computer, these few scientist-polygraphers have painted a picture of a brave new world of science fiction that many find appealing. And although their own research demonstrated that college students could be taught to "beat" the lie detector in a half-hour's instruction, these scientist-advocates apparently do not believe that professional spies and criminals could learn the same lesson.

In this book I have tried to present the relevant facts and arguments in a form accessible to the general reader rather than to ask anyone to rely on my opinion. After all, as the saying goes, we are not dealing here with rocket science. I believe that a British or a European audience, not infected from childhood with the myth of instrumental lie detection, would agree that I have made my case. To such an audience, the burden of proof would be on the person who claims that a polygrapher can, in two hours, determine the veracity of a witness, or the guilt or innocence of a criminal suspect, with almost perfect accuracy. But, in the United States, this burden has oddly shifted to the shoulders of the critic. I have described the kinds of field studies that would be required to give a dispositive empirical answer to the question of lie detector accuracy, but without the cooperation of the fraternity of polygraphers, these studies are unlikely to be conducted.

A 1993 U.S. Supreme Court opinion, *Daubert v. Merrell Dow Pharmaceuticals*,[1] while it did not deal directly with the polygraph, has changed the rule for admissibility of scientific evidence at trial in federal courts and in state courts that follow the federal lead. In many jurisdictions where polygraph results were formerly inadmissible per se, courts are now holding evidentiary hearings for the purpose of determining whether the polygraph results submitted meet the new *Daubert* criteria of scientific validity. The U.S. Court of Appeals of the Armed Forces, in 1997, decreed that judges in military courts-martial, when polygraph evidence is proffered, must do the same. To assist judges and lawyers in meeting these new *Daubert* requirements, a major legal publishing house commissioned a handbook with chapters on new or contentious scientific evidence, ranging from DNA "fingerprinting" to eyewitness identification, each written by appropriate experts.[2] Because of its especially contentious nature, the polygraph issue was assigned two chapters, one by its scientist-advocates, D. Raskin, C. Honts, and J. Kircher, and the other by my colleague, W. G. Iacono, and myself. Judges seeking guidance on the polygraph issue must therefore choose (as, indeed, they do daily in their courtrooms) between two diametrically opposed sets of arguments.

All five of these authors are members of the Society for Psychophysiological Research (SPR), the principal international association of scientists engaged in research on brain waves, cardiovascular reactions, palmar sweating, and other physiological responses that can reflect covert mental or emotional events and thus provide a (somewhat clouded) window on the mind. Polygraphic lie detection can be regarded as an example of applied (or misapplied) psychophysiology. Judges weighing the merits of these two conflicting chapters might be influenced by the fact that Dr. Iacono and I are both past presidents of SPR, but, by the same token, they might also give weight to the results of two polls of the opinions of SPR members concerning the polygraph.

Two Prior Polls

Members of SPR—among whom are a number of practicing polygraphers and the editors of the trade journal *Polygraph*—were surveyed by telephone in 1984 by the Gallup Organization on behalf of a litigant seeking to have a lie test result admitted at trial. A similar survey was conducted by mail in 1993 by a student of Professor Honts at the University of North Dakota.[3] The answer to one question asked in both surveys has been used to suggest acceptance of polygraphy by the scientific community. Respondents were asked to choose one of four statements that best described their "opinion of polygraph test interpretations" to determine "whether a subject is or is not telling the truth." The responses this question received in both prior surveys are reproduced in Table 12.1.

Raskin, Honts, and Kircher drew special attention in their chapter to the fact that about 60% of respondents in both prior surveys endorsed

Table 12.1. Three Surveys of the Opinions
of Members of the Society for Psychophysiological Research
Regarding the "Usefulness" of Polygraph Test Interpretation

Question: Which one of these four statements best describes your own opinion of polygraph test interpretations by those who have received systematic training in the technique, when they are called upon to interpret whether a subject is or is not telling the truth? It is—	Gallup 1984[a] (N = 152)	Amato 1993[b] (N = 135)	Iacono & Lykken 1994–95[c] (N = 183)
A. Sufficiently reliable to be the sole determinant.	1%	1%	0%
B. A useful tool when considered with other available information.	62%	60%	44%
Between "B" and "C"[d]	2%	0%	2%
C. Of questionable usefulness, entitled to little weight.	33%	37%	53%
D. Of no usefulness.	3%	2%	2%

[a]Telephone survey.
[b]S. L. Amato and C. R. Honts, What do psychophysiologists think about polygraph tests? A survey of the membership of SPR, *Psychophysiology*, 1993, *31*, S22 (Abstract).
[c]This poll was conducted by D. K. Rasmussen for her *summa* thesis under the supervision of Professors Iacono and Lykken.
[d]Although not offered as an option, in two of the surveys respondents indicated a choice that fell between alternatives "B" and "C."

alternative "B" indicating that polygraph interpretation is "a useful [diagnostic] tool." This finding led advocates to conclude that the membership of SPR considers polygraph tests "useful for legal proceedings"[4] even though the question does not ask for an opinion about the use of polygraph tests in court. Indeed, the meaning of endorsements of this alternative appears ambiguous. There is general agreement that polygraph testing can be useful as an interrogation tool, and respondents who chose option "B" may have interpreted the statement to mean nothing more than this. Moreover, the prior surveys made no distinction between the CQT, or lie detection, which is controversial, and another polygraphic interrogation method, the guilty knowledge test, or GKT, which many consider to be scientifically credible. It is possible that some SPR members answered this central question in both surveys with the GKT, rather than lie detection, in mind.

Two Recent Polls

These results seemed limited in scope and unclear in their implications. Moreover, a response rate of only 30% had been achieved in the previous mail poll, and according to the former U.S. Office of Statistical Standards, significant caution is recommended when response rates drop below 50%.[5] Consequently, Bill Iacono and I, with the capable assistance of D. K. Rasmussen, endeavored to assess more thoroughly scientists' opinions about polygraphy by (1) covering a variety of issues that are controversial in the literature, (2) obtaining opinions from the members of two relevant scientific organizations, and (3) structuring our surveys to optimize the likelihood of a high response rate.

In our questionnaires, to avoid the vagueness of prior surveys, we provided respondents with descriptions and definitions of the various *methods* (described without commentary) using primary source material wherever feasible. Our purpose, in short, was to ensure that the scientists surveyed were all provided with the same accurate, complete, and unbiased descriptions of the methods and issues about which they were being asked to give their scientific opinions. Information that was provided to the survey respondents is reproduced in the section that follows.[6]

Information Provided to Respondents

The Nature of the Control Question Test   The CQT compares the physiological disturbance caused by relevant questions about the crime (e.g., for the O. J. Simpson case, "On June 12, did you stab your ex-wife, Nicole?") with the disturbance caused by "control" questions relating to possible prior misdeeds (e.g., "Before 1992, did you ever lie to get out of trouble?"). As characterized by Raskin[7]:

> the control questions deal with acts similar to the issue under investigation but are more general in nature. They cover many years in the prior life of the subject and are deliberately vague. Almost anyone would have trouble answering them truthfully with a simple "No" ... Control questions are designed to provide the innocent suspect the opportunity to become more concerned about questions other than the relevant questions and to produce stronger physiological reactions to the control questions. If the subject shows stronger physiological reactions to the control as compared to the relevant questions, the test

outcome is interpreted as truthful. Stronger reactions to the relevant
questions indicate deception.[7]

According to Dr. Raskin,[8] a typical introduction of control questions by the
examiner for a case of theft of a ring would be:

> "Since this is a matter of a theft, I need to ask you some general questions
> about yourself in order to assess your basic honesty and trustworthiness. I
> need to make sure that you have never done anything of a similar nature in the
> past and that you are not the type of person who would do something like
> stealing that ring and then would lie about it.... So if I ask you 'Before the age
> of 23, did you ever lie to get out of trouble?' you could answer that 'No'
> couldn't you?" Most subjects initially answer "No" to the control ques-
> tions. If the subjects answer "Yes," the examiner asks for an explana-
> tion ... [and] leads the subject to believe that admissions will cause the
> examiner to form the opinion that the subject is dishonest and there-
> fore guilty. This discourages admissions and maximizes the likelihood
> that the negative answer is untruthful. However, the manner of intro-
> ducing and explaining the control questions also causes the subject to
> believe that deceptive answers to them will result in strong physiologi-
> cal reactions during the test and will lead the examiner to conclude
> that the subject was deceptive with respect to the relevant issues
> concerning the theft. In fact, the converse is true. Stronger reactions to
> the control questions will be interpreted as indicating that the subject's
> denials to the relevant questions are truthful.

Countermeasures   In laboratory research by Honts, Raskin, and colleagues,[9] "guilty" subjects were trained in the use of "countermeasures" to
be applied while control questions were being presented during their
examinations. The actual training consisted of instructing subjects

> to press their toes to the floor, to bite their tongue, or ... to count
> backward by 7s from a number larger than 200 when the control
> questions were asked. Each countermeasure subject was instructed to
> begin the countermeasure as soon as he or she recognized a control
> question, stop just long enough to answer, and then continue the
> countermeasure until the next question began. Each countermeasure
> subject was then read a set of questions from a typical CQT and was
> coached in using his or her countermeasure unobtrusively so that it
> would not be detected by the polygraph examiner during the subse-
> quent test. None of the questions used in this practice test was used in
> the actual polygraph examinations, and subjects were not informed of
> the order of the questions during the examination. The counter-
> measure training required a maximum of 30 min.

The countermeasures training did not involve attaching subjects to a
polygraph to give them the opportunity to learn how their counter-

measure maneuvers affected their physiological recording. The results indicated that

> the mental and physical countermeasures were equally effective: Each enable approximately 50% of the Ss to defeat (i.e., appear truthful on) the polygraph test.... Moreover, the countermeasures were difficult to detect either instrumentally [i.e., by inspecting the physiological records] or through observation.

The Directed Lie Test (DLT)   As an alternative to the control questions that are used in the CQT, some examiners substitute a "directed lie" question for the control question. An example would be "Have you ever told a lie?" or "Have you ever broken a rule?" to which subjects are told to answer "No." Subjects are also instructed to think about a particular time they told a lie or broke a rule when they are asked these questions. Directed lie questions are explained to the subject by the examiner as follows:

> I need to have questions to which you and I both know you are lying. That way, I can be sure that you continue to respond appropriately when you are lying and that you remain a suitable subject throughout this test. Raskin goes on to note that: guilty subjects should show stronger reactions to the relevant questions. However, subjects who are truthful in response to the relevant questions will be most concerned that the "appropriateness" of their reactions to the directed lie questions will show that they are suitable subjects and will demonstrate that their reactions are different when they are truthful. This focus of concern should enhance the reactions of truthful subjects to the directed lie questions, making them stronger than reactions to the relevant questions.[10]

The Guilty Knowledge Test   The GKT attempts to detect, not lying, but whether the subject possess "guilty knowledge," that is, knowledge that only the perpetrator of the crime and the police investigators would have. For example, "If you were at the crime scene, Mr. Simpson, you would know what Nicole was wearing. Was she wearing a green swim suit? —a black cocktail dress? —a white tennis outfit? —a red blouse and slacks? —a blue bathrobe? —a T-shirt and jeans?" A GKT might consist of ten such items. Guilt would be suggested by a consistently stronger response to the correct guilty knowledge alternative among these items. Although the GKT is seldom used in the field, it has been found to be a highly accurate procedure in most laboratory studies.

How the Surveys Were Conducted

To encourage responding, we limited the length of the survey to three pages of questions. To broaden the coverage of CQT-related issues, we used three different questionnaires with overlapping items. As did Gallup and Amato, we surveyed members of SPR. In addition, however, because it is the psychological and psychometric aspects of polygraphy that are the focus of controversy rather than instrumental or physiological issues, we also surveyed a group of scientists who are especially knowledgeable in these areas, namely, those psychologists who have been distinguished by election as Fellows of Division One (General Psychology) of the American Psychological Association. In the SPR survey we repeated the question reproduced in Table 12.1 that had been used in prior surveys, but we also included a broad range of questions designed to provide a more detailed, more specific, and less ambiguous record of the opinions of these representatives of the two relevant scientific communities.

The surveys were sent to a randomly selected 50% of all nonstudent members of SPR, and also to all listed Fellows of APA's General Psychology Division.[11] Only those members of either group with U.S. addresses were solicited to participate. By providing information about the various methods in the words of the techniques' leading advocates, we sought to insure that the descriptions would be accurate and presented in the most favorable light. Respondents were instructed to answer questions based on what they "knew and believed" about polygraph testing.

Results

Of the 214 SPR members who received surveys, 195, or 91%, returned questionnaires. In the case of APA Fellows, the pool consisted of 226 individuals and 168, or 74%, returned usable questionnaires. This latter response rate, while lower than that obtained from SPR members, is still considered high for mail survey research.[12]

The SPR Survey   We asked our SPR respondents to answer the same question used in the Gallup and Amato and Honts surveys, but we made it clear that it referred to conventional "lie detection" with the CQT. The results, shown in the right-hand column of Table 12.1, indicate that a majority of psychophysiologists consider that the usefulness of the CQT is "questionable." Nevertheless, 44% of our respondents thought that the technique might be a "useful [diagnostic] tool," so we are left to wonder whether nearly half of these experts actually believe that the CQT is valid as a test, even valid enough to be introduced as evidence at trial. To address these possibilities, respondents were asked the questions listed in Tables 12.2 and 12.3.

Turning to Table 12.2, only 36% considered the CQT to be based on principles that are "scientifically sound." In contrast, 77% indicated that the GKT had a sound scientific basis. Although questions of admissibility are decided not by scientists nor polls, but by judges, and on legal rather than scientific grounds, lie detector advocates had interpreted the earlier survey results to mean that SPR members would favor such admissibility. Therefore, we asked our respondents explicitly whether they would favor the introduction of CQT findings as evidence in court. About three-fourths of respondents were opposed to such use of the lie detector.

Responses to the remaining SPR survey questions are summarized in Table 12.3. Polygraph proponents uniformly assert that the CQT is over 85% accurate. To determine how such claims were viewed by the SPR membership, we asked subjects the following question: "Proponents of

Table 12.2. Responses to Questions about Polygraphy Asked of Members of the Society for Psychophysiological Research and Fellows of APA Division One

Questions	"Yes" responses from:	
	SPR members	APA fellows
Would you say that [the CQT, GKT, DLT] is based on scientifically sound psychological principles or theory?		
CQT	36%	30%
GKT	77%	72%
DLT[a]		22%
Would you advocate that courts admit into evidence the outcome of a CQT, that is, permit the polygraph examiner to testify before a jury that in his/her opinion, either the defendant was [deceptive or truthful] when denying guilt?		
Deceptive	24%	20%
Truthful	27%	24%

[a]Question asked only of APA Fellows.

Table 12.3. Opinions about Polygraphy
by Members of the Society for Psychophysiological Research

Questions	Agree	Disagree
1. The CQT is at least 85% accurate		
a. In tests of guilty suspects.	27%	73%
b. In tests of innocent suspects.	22%	78%
2. "Friendly" CQTs are more likely to be passed than those taken under adversarial conditions.	75%	25%
3. The CQT can be beaten by augmenting one's responses to the control questions.	99%	1%
4. It would be reasonable for courts to give "substantial weight" to results of laboratory studies to estimate CQT validity in real life.	17%	83%

polygraphy typically assert that the CQT is highly accurate, with hit rates of 85% or better for both guilty and innocent suspects. Based on what you know and believe, to what extent would you agree that the CQT is accurate at least 85% of the time in real-life applications for guilty and innocent suspects?" For both the guilty and the innocent, about three-quarters of SPR members disagreed that the CQT is accurate at least 85% of the time.

To tap member opinion about friendly tests, we instructed respondents to assume that a defendant awaiting trial plans to take either a private CQT from a polygrapher hired by his lawyer, or an adversarial CQT from a police examiner, and that the polygraph examiners had equivalent training, experience, and expertise. Seventy-five percent of SPR members thought that the friendly test was more likely to be passed.

As the next item in Table 12.3 indicates, 99% of respondents accept the notion that "the CQT can be beaten by augmenting one's responses to the control questions." Finally, SPR respondents were asked the following question: "Some empirical studies of the CQT involve laboratory manipulations in which those experimental subjects assigned to the guilty group are instructed to commit a mock crime (e.g., to steal a ring from a desk) and then submit to a polygraph test. In your opinion, is it reasonable for judicial proceedings to give substantial weight to the polygraph 'hit rates' obtained in these laboratory simulations to estimate the validity of the CQT for innocent and guilty suspects in real-life criminal investigations?" Only 17% agreed that this was reasonable.

The Survey of Division One Fellows   The opinions of our survey of distinguished general psychologists are summarized in Tables 12.2 and 12.4. As Table 12.2 indicates, when APA Fellows were asked the same questions about the soundness of the CQT and the GKT and the advisability of admitting polygraph evidence in court that the SPR members were asked, they responded in a similar way. Only APA members were queried about the soundness of the assumptions of the directed lie test; only 22% thought them to be scientifically sound.

Asked to make their "best estimate of the accuracy of the CQT in testing both innocent and guilty criminal suspects" by picking a number from 0 to 100%, Fellows estimated the accuracy to be around 60% for both innocent and guilty suspects (Table 12.4). Nearly three-fourths said they either "definitely would" or would be "inclined to" take a privately administered CQT if they were personally guilty of some crime and wished to use the test results to deflect suspicion from themselves. Interestingly, when instructed to assume they were innocent with the opportunity to take a police-supervised CQT, the results of which would be presented to a jury, only 35% would be "inclined to" take it.

Over 90% felt that "professional criminals, defendants with unscrupulous lawyers, or foreign agents" could learn to beat the CQT using countermeasures, and 75% expressed "moderate" to "high" confidence

Table 12.4. Opinions about Polygraphy by Fellows of Division One (General Psychology) of the American Psychological Association[a]

1. What is your best estimate of the accuracy of the CQT?	
a. For testing innocent suspects:	63.0
b. For testing guilty suspects:	59.5
2. If guilty, would you take a "friendly" CQT?	75
3. If innocent, would you take an adversarial CQT?	35
4. Are criminals and spies likely to beat the CQT?	92
5. Are you confident that you could personally beat a CQT?	75
6. Can the CQT accurately be called a standardized procedure?	20
7. Is the CQT relatively independent of differences among examiners in skill and subjective judgment?	10
8. Strong empirical evidence would be required before accuracy claims of CQT proponents could be believed.[b]	93

Note. All numbers are percentages.
[a]In Questions 2–8, numbers indicate percentage of "Yes" answers.
[b]Question asked only of the APA Division One Fellows.

that they could learn to use effective countermeasures "well enough to defeat a CQT." Regarding the CQT's psychometric soundness, it can be seen that only 20% thought the CQT could be accurately described as a standardized test, and only 10% thought it to be an objective procedure, "relatively independent of differences among examiners in skill and subjective judgment." Most (93%) agreed with the following statement: "Strong and unequivocal scientific evidence of validity should be required before the accuracy claims of polygraph proponents are believed."

Summary

Due to a 1993 U.S. Supreme Court decision, federal and probably many state court judges are being asked to introduce the results of CQTs into evidence at trial. To provide these courts with the assistance of the relevant scientific communities in making these important rulings, we have surveyed the opinions of two groups of scientists, members of SPR and also general psychologists distinguished by election as Fellows of APA Division One. The two groups, quite different in terms of the background and expertise of their members, nevertheless gave similar answers to the questions that were common to both surveys (see Table 12.2), so we can reasonably assume agreement on the questions unique to each of the two questionnaires.

Both of our surveys, which elicited response rates that are high for mailed questionnaires, produced results that complement each other and stand in contrast to prior surveys on the topic. Our respondents do not accept the accuracy claims of the polygraph community. Three-fourths of the SPR members thought it unlikely that the validity of the CQT could be as high as 85% and the APA Fellows estimated its average validity at about 61% (versus a chance expectancy of 50%). They do not believe that CQT results should be admitted as evidence in court. Seventy-four percent of SPR members and 78% of Division One Fellows said that they would oppose the use of such evidence. Nor do SPR members agree that laboratory studies with mock crimes should be used by courts to estimate CQT validity in real life. They do not accept friendly and adversarial tests as equivalent. By a three-to-one ratio, they agreed that guilty subjects are likely to pass CQTs taken under friendly rather than adversarial circumstances. Nearly three-fourths of the Fellows would take a friendly test if guilty, but only about a third would take an adversarial test if innocent.

Over 90% of both groups agreed that they believe that both criminals and they themselves could learn countermeasures to defeat the CQT. Their confidence in the DLT is no higher than their faith in the CQT, which they characterize as neither standardized nor objective.

It was not the case that members of these two organizations were negatively disposed toward polygraphic interrogation generally. In contrast to their concerns about the soundness of the CQT, which only about one-third endorsed, almost three-fourths of respondents viewed the GKT as scientifically sound. Since we did not ask the question, we cannot claim that these experts would advocate the admissibility at trial of the results of guilty knowledge tests. Although it is my personal belief that the GKT is a promising forensic tool, I would not myself advocate its admissibility until considerably more experience by law enforcement with this technique has been accumulated.

Why our SPR survey elicited a different response to the question in Table 12.1 than those of Gallup and Amato is not clear. As noted, unlike the prior surveys, it was our intention to make clear which survey questions dealt only with the CQT. Also, our sample may be more representative of the SPR membership than those of the other surveys. Of the SPR members we solicited, 86% answered this question, compared with only 30% of those polled by Amato. No information was provided by the Gallup Organization about how many SPR members called in 1984 were willing to cooperate. The fact that the Gallup survey preceded ours by more than ten years may also be a factor in accounting for the differential response rates.

None of our respondents could recall participating in the Gallup survey and only 8% indicated that they had responded to the Amato survey. This low degree of overlap is difficult to explain, but it suggests that those who were previously surveyed may not have been a representative subset of SPR members.

While I have tried in this book to present an account of polygraphy that is complete and clear enough to permit the reader to form an independent judgment, it would be only natural to wonder what professional scientists, experts in psychophysiology or in psychological assessment, would make of the same evidence. These two new surveys provide that information. Although an exceptional majority of our respondents expressed a poor opinion of the lie detector, you might continue to wonder how nearly one-third of these scientists could say that the CQT "is based on scientifically sound psychological principles or theory" or how nearly a fourth of them could suppose that it is accurate as much as 85% of the time. I confess to some wonderment on this score myself. However, after more than 40 years as a professional psychologist, I have formulated a principle that I believe may provide an answer to this puzzle and others. It is my impression that, in most controversies, whether scientific, ethical, or political, about 25% of the population tend to be right most of the time, another 25% tend always to be wrong, while the remaining 50% are the swing votes that will get it right eventually if one can induce them to pay attention. I have seen no reason to believe that scientists are any less obedient to this principle than are people in general.

Chapter 13

THE TOOLS OF DIOGENES: AN OVERVIEW

We have reviewed the principal lie detection techniques currently in use (Table 13.1). I should make it clear that numerous other lie test formats have been expounded in the literature, many with exotic names such as The Matte Quadri-Track Zone Comparison Technique, The DoDPI Bi-Spot and Tri-Spot Techniques, The General Question Test, The S-K-Y Test, and so on. In a recent text,[1] a polygrapher named J. A. Matte lists a total of 18, but they are all minor variations on these same themes, involving in diverse combinations the same implausible assumptions.

We have seen that no wholly satisfactory assessment of the validity of any of these methods has as yet been done. Such an assessment would require some large police agency to test every possible suspect over a period of months or years, leaving the tests unscored so that they cannot influence the subsequent investigations, and omitting any posttest interrogation of the suspects by the polygraph examiners. Once any case has been cleared, yielding certain knowledge of which suspects had been truthful and which deceptive, knowledge uncontaminated by the polygraph results, the tests of those suspects would then be scored by polygraphers ignorant of the case facts. Because the existing field studies used polygraph-induced confessions as the criteria for ground truth, we know that the results of these studies provide only upper-limit estimates of lie detector validity. These upper-limit estimates suggest an accuracy of about 60% in detecting truthful responding and about 85% in detecting lying (Table 8.2) versus a chance value of 50%. It is certain that these estimates would be revised downward based on the results of an adequate field

Table 13.1. The Main Types of Polygraphic Lie Tests

Name of test	Originator	Subject of "control" questions	Comments
Relevant/Irrelevant (R/I) General Series Test	Larson	Irrelevant, innocuous matters.	The earliest method, yields many false-positives.
Control Question Test (CQT)	Reid	Past misdeeds similar to but less serious than crime in question; answers are assumed lies.	Standard method for specific-issue work. Only method tested for validity; 64%–72% average, half of truthful subjects fail.
Zone of Comparison (ZOC)	Backster	Same as CQT.	Variant of CQT; similar assumptions, equivalent validity.
Directed Lie Test (DLT)	Sheila Reed at DoDPI[a]	Subject instructed to deny or "lie" about commonplace sins.	Variant of CQT, no good field data on validity. Assumes forced "lie" as arousing as a real lie.
Truth Control Test (TCT)	Reid	Fictitious crime; subject led to think he is equally suspect, equally at risk.	Seldom used, never tested for validity. Requires elaborate deception of suspect.
Positive Control Test (PCT)	Reali	Subject required to answer each question with a lie, then truth; lie is the control.	No validity data. Assumes that forced "lie" as arousing as real lie, forced truth equivalent to volunteered truth.
Screening Test	(evolved)	Only relevant questions; serve as controls for each other.	Standard employee screening test. Widely used; no validity data.
Peak of Tension (POT)	Keeler	Like a multiple-choice item; incorrect alternatives are the controls.	Only examiner and guilty suspect know correct alternatives. Several POT items yield a Guilty Knowledge Test.
Searching Peak of Tension (SPOT)	Keeler	As above but examiner does not know correct alternative.	Used only if indications can be confirmed; a search procedure rather than a test.

Note. Unless the test conclusion is based solely on objective scoring of the polygraph charts, any of these methods degenerates to a clinical interview, its validity determined by the intuitive skills of the examiner. Typical "human lie detectors" seem to be wrong 30% to 50% of the time. Polygraphic lie detectors do no better.
[a]Department of Defense Polygraph Institute at Ft. McClellan, Alabama.

study in which ground truth was established wholly independently of the test results.

We shall see in Chapter 19 that is relatively easy to learn to beat any of the control question lie tests through the use of covert self-stimulation after each control question is presented. Because the details of these counter-measures are now in the public domain, we must assume that they will be increasingly widely known and employed by at least the more sophisti-cated criminals, foreign agents, and other malefactors. Thus, if an ade-quate validity study were to show that the true sensitivity of the polygraph test for lying is, say, 70%, rather than 85%, based on tests of unsophisti-cated suspects, then we should have to assume that tests of sophisticated guilty suspects, including criminals able to afford sophisticated legal rep-resentation, would tend to be more often wrong than right.

Because of this lack of really adequate empirical validation of the lie detection methods, we have systematically explored the opinions of scien-tists most knowledgeable about psychophysiological techniques and about the general problems of mental measurement. The great prepon-derance of these scientists agreed with the views expressed here, namely, that the assumptions of lie detection and the validity claims of its propo-nents are implausible.

Lies, Lies, Lies!

One important point about the various lie detection methods that we have only touched on in passing deserves explicit emphasis in this sum-ming up. *All of these techniques fundamentally depend on deception*—not just in one way and not just in little ways. The theory and assumptions of polygraphic interrogation require the examiner to successfully deceive each subject that he tests in several basic ways. First, he must persuade the subject that being untruthful or even unsure about his answers to the control questions may cause him to fail the test, although in fact the opposite of this is true. Second, when he administers the "stim" test in order to impress the subject with the accuracy of the technique, the exam-iner has two choices, both of them deceptive. He can use the original Reid "pick-a-card" method in which the deck is either stacked or marked so that the examiner can be sure to guess the right card. Alternatively, he can use the Raskin "pick-a-number" method in which he deceitfully explains that he is "determining what your polygraphic response looks like when you lie." The truth is, of course, that individuals do not show characteristic physiological response patterns when they lie that they do not also show when telling the truth. Third, throughout his interactions with the subject, the examiner must convey an impression of virtual infallibility. The stimtest is just a component of this basic deception. The purpose is benign enough; if guilty subjects are convinced the polygraph will reveal their guilt, then they are more likely to respond strongly to the relevant questions. If innocent subjects are similarly convinced, then they will tend not to respond so strongly. Moreover, because most examiners truly believe in their near-infallibility, because as we have seen they are the victims of their own deceptive art, they may convey this needed impression not only effectively but also without conscious guile. Nonetheless, the polygraph test, as we have seen, has an accuracy closer to chance than to infallibility; the innocent suspect being tested by the police faces worse odds than in a game of Russian roulette. The fact that most polygraph examiners are not aware of these facts (indeed, they may be the last to know) is not an adequate excuse. Fourth, when the subject is interrogated after a polygraph test, he may be the victim of repeated deceptions. "This unbiased, scientific instrument is saying that you're not telling the truth about this, John!" "Why don't you tell me whatever it is that you feel guilty about,

Figure 13.1. This cartoon by Michael Wille, which first appeared in *The Atlantic Monthly* (© 1984, used with permission), confirms the aphorism that a picture is worth a thousand words.

Mary, then maybe you will do better on the next test." "With this polygraph chart, George, no one is going to believe you now. The best thing you can do is to confess and make the best deal you can."

I will confess here that I do not personally object to certain harmless deceptions of criminal suspects that might lead to verifiable confessions and a quick and easy solution to a criminal investigation. But a procedure that claims to be a genuine test for truth that cannot hope to succeed even by its own theory and assumptions unless the subject is successfully deceived in certain standard ways is an invitation to abuse, abuse by examiners and especially by sophisticated criminals and spies. I submit that it is madness for courts or federal police and security agencies to rely on polygraph results for this reason alone. As we have seen, of course, there are many other reasons for this same diagnosis.

Part III

LIE DETECTION:
THE APPLICATIONS

(Interview with Roc, age seven): What happens when you
tell lies? *You get punished.* And if you didn't get punished,
would it be naughty to tell them? *No.* I'm going to tell you
two stories. There were two kiddies and they broke a cup
each. The first says it wasn't him. His mother believes him
and doesn't punish him. The second one also says it
wasn't him. But his mother doesn't believe him and
punishes him. Are both lies they told equally naughty? *No.*
Which is the naughtiest? *The one who was punished.*

—JEAN PIAGET, *The Moral Judgement of the Child*

*Listen, I don't know anything about polygraphs and I don't
know how accurate they are, but I know they'll scare the hell
out of people.*

—PRESIDENT RICHARD M. NIXON, Oval Office Tape,
July 14, 1971

This section begins with an examination of the widespread but seldom
discussed practice of modern American business to do its own police
work. Many American businesses, large and small, deal with intramural
crime by hiring private investigators or by maintaining their own security
operatives. Prior to the federal Employee Polygraph Protection Act of 1988,

these private investigators leaned heavily on the polygraph as a chief tool in their work, and due to loopholes in the Act, many still do. As Chapter 14 makes clear, this private police work is frequently effective, but any civil libertarian will find other consequences of this trend disturbing.

By far the most common use of lie detection is in the screening of job applicants and the periodic screening of current employees, by the FBI, by federal security agencies, and by many state and local police departments. Chapter 15 discusses the arguments for and against this practice. A related issue is the use of paper-and-pencil tests, questionnaires, that are alleged to measure "honesty" or "trustworthiness." These developments are critically reviewed in Chapter 16. Chapter 17 is devoted to what I believe is the most effective feature of the polygraph examination, its indisputable ability to elicit damaging admissions or confessions, at least some of which prove to be false.

In recent years, the legal barriers against the use of lie test results as evidence in court proceedings have weakened. In an increasing number of U.S. jurisdictions, polygraphers who have tested witnesses or the defendant himself are being accredited as expert witnesses and permitted to offer their opinions as to the credibility of the person tested. An analysis of this difficult and important problem will be found in Chapter 18.

Chapter 19, entitled "How to Beat the Lie Detector," is not really intended as a handbook for hoodlums. I explain earlier (in Part II) why I think that the lie detector tests do so poorly in identifying truthful responses, and Chapter 19 explains how sophisticated criminals as well as foreign agents can prevent the polygraph from detecting deception. But this chapter is primarily concerned with my personal evaluation of the probable effects on American society of the growing uses of modern truth technology and my own recommendations about what should be done. If, as I think, the lie detector can be beaten by a guilty criminal who has been schooled by some knowledgeable Fagin or unscrupulous lawyer (or who has read this book), then this is just an additional reason why we should consider legislation and other ways to "beat"—or to beat back—the metastasizing of the lie detector throughout the tissues of our social system.

Chapter 14

TRUTH, LTD.:
THE POLYGRAPHER
AS PRIVATE DETECTIVE

PAUL ALTMEYER: *"Do you think an employer should rely solely on the results of one of these tests?"*
JOHN REID: *"Well, true, I suppose they could put a person back to work and say, 'Now don't do that again.'"*
ALTMEYER: *"Well, you're the judge and jury then?"*
REID: *"Well, true in that particular thing, except that the penalty is just losing your job ..."*

—NBC News Interview with John Reid, 1977

Someone must have been telling lies about Joseph K. for without having done anything wrong he was arrested one fine morning.

—F. KAFKA, *The Trial*

Early in 1986 a major U.S. television network, CBS, conducted its own test of the polygraph. CBS owns the magazine *Popular Photography* and arranged for that subsidiary's office manager to contact four different polygraph firms in New York City, engaging their help in determining which of four employees had stolen a valuable camera. As the polygraphers arrived (on different days) to do the testing, each one was advised that, while all four employees had had access to the camera, the manager's suspicions were focused on one individual; a different individual was specified for each polygrapher.

No camera was missing, in fact, and the four employees, who were privy to the scheme but did not know whom the manager had singled out, were instructed merely to truthfully deny that they had stolen any camera. They were each offered $50 if they succeeded in passing the polygraph test. Each of the four polygraphers positively identified a culprit; in each case, the innocent employee thus stigmatized as a thief and a liar was the person whom the polygrapher had been told was especially suspect. All four polygraphers expressed total confidence in their diagnoses and one of them filmed, covertly, explaining to the manager that he should "find some other excuse for firing this person; it's illegal in New York State to fire somebody solely because of a polygraph test."

Intramural Crime

Employee crime—pilfering, theft of merchandise, tools, or equipment, embezzlement, sabotage, and vandalism—is an epidemic problem in the United States. Annual losses to American business are counted in the billions of dollars. "Epidemic"—*in the people*—is particularly appropriate because not only do the people in general pay for these losses, but, it would appear, people in general are responsible for them. Under the stress of preemployment polygraph testing, as many as 75% of job applicants admit to some sort of thievery from previous employers.[1] In view of the tax monies paid out for police protection, it comes as a surprise to learn that official police agencies are never even notified in most of these affairs. Increasingly, business is doing its own police work, hiring security personnel or engaging the services of private detective agencies. We are seeing a return to the conditions of the Middle Ages when the nobility and the rich merchants, represented now by the corporation and the private company, provided their own guards and police, small private armies in some instances, to protect themselves and their property. The reasons for this trend, like most business decisions, are pragmatic. If the corporation is experiencing unexplained losses, if someone is sabotaging the assembly line, if some substantial theft bears the earmarks of an "inside job," the official police are unlikely to provide efficient and effective help. The company wants the losses stopped, wants the culprit identified so that he can make restitution if possible and, in any case, be eliminated from the payroll and from the plant. So, the private police are called in, and more than likely, they will bring with them a polygrapher.

This private detecting is frequently successful, which is precisely why it is increasingly utilized. Let us look at it in action. In 1976, a commercial bakery in Southern California was facing ruin, its 125 employees about to lose their jobs. Outraged retailers were returning bread shipments, loaves found to contain shards of glass, bits of barbed wire. Frantic checking of the raw materials, the mixing and baking processes, made it plain that this could be no accident caused by faulty equipment or simple carelessness. It was sabotage, the work of someone deliberately contaminating the dough. After consultation with the employees' union, the bakery called in Intercept, a private polygraphy firm in Hollywood. Two examiners traveled to the bakery, portable polygraphs in hand, and began testing the employees. The eighth person tested, a man of 18 years' experience, confessed in the course of the examination that he had been responsible for all the damages. Angered because someone else had received a promotion to which he thought he was entitled, this man had brooded for two years and then began his revenge. The police had been no help at all. What could they have done? The dozen employees with easiest access to the mixing vats might have been interrogated officially, their background searched for some motive, their families and friends questioned. Rigorous security measures might have been imposed to restrict access to the mix, to require pocketless work clothes, and so on. A smart policeman might have discovered that two-year-old resentment and followed up the clue. Just having a police detective prowling around, asking questions, might have produced results. But the two polygraphers *did* produce conclusive results within 24 hours, and at relatively little cost.

During November of 1978, a gasoline service station in Salt Lake City experienced the loss of about $700, primarily in cash. After $290 disappeared from the till in one night, Polygraph Screening Service of that city was engaged to test the employees. During the pretest interview, a young man hired only two weeks previously admitted pocketing an average of $30 to $40 a day in sums collected for service or gasoline and never rung up. After a polygraph test of the CQT type on which he showed considerable reaction to the questions concerning the $290 theft, he said that he had inadvertently left the key in the cash box that night while visiting the toilet and found the money missing when he returned. Thus, while continuing to deny the major theft, he admitted stealing an aggregate of several hundred dollars over a period of two weeks, losses that stopped when he was fired. He could have been caught, perhaps, by surveillance or by having professional "shoppers" give him the exact change for a purchase while a confederate watched to see what he did with the money. The city police probably would not have helped. As in the previous example, the polygrapher solved the problem expeditiously and by inducing a confession. The only use made of the examiner's diagnosis that this young man was lying about the major theft was that suspicion was diverted from the other employees.

In some instances, polygraph testing not only catches the culprit, but also exonerates innocent employees whom management is prepared to discharge on the basis of suspicion alone. An art dealer in Beverly Hills was losing trade secrets—names of clients and contacts—to a competitor and had accused a new assistant. A polygrapher cleared this assistant (another case of the CQT producing the sort of truthful chart that one would seldom expect) and then proceeded to obtain a confession from a long-time, trusted employee. A theater manager, citing losses, recommended that all the cashiers, concession clerks, and ushers be polygraphed. The examiner concluded that none of them were responsible for the losses and recommended that the manager himself be asked to take the test. Thereupon the manager confessed a plan to cover up his own thievery by shifting attention to his part-time help. I have no evidence that all of these young people "passed" their tests in the sense that all of their charts would be blindly scored as truthful. But at least the examiner was not persuaded that he had found his thief until he got to the manager. Scores of similar examples could no doubt be accumulated by polling private polygraphers around the country. The cases I have cited were obtained from two firms run by men whom I know and trust.

But these successes must be balanced against scores of examples (like the Coker case below), where employees were victimized in a kangaroo court with a machine serving as the jury.

The Coker Case

Extracts from a letter dated February 17, 1975:[2]

"My spirit is broken and my mind weary, but I am going to do my best to tell you just what happened, out of love for my wonderful husband and the hope that you can help him. He has always been a quiet, easy-going, humble person and the shock of his being blamed for something he is not guilty of has put him in a state of depression that worries me every moment. I am trying to let him know how much I care, know without a shadow of a doubt he is innocent, and how very much the children and I need him. So far, none of my assurance has seemed to help very much.

"Sometime in December, 1974, Willis Foster came to my husband and told him he had been sent from Charleston to find out what was going on in the store. The Lake City Piggly Wiggly, where Mack was the assistant manager, had been having some shortage problems before, but in the past two quarters, they said the shortage reached a peak of $13,000 one quarter and $10,000 the last quarter. Near the end of January, Mr. Lewis, the manager, was called to Charleston and returned with the news that the people from Truth Associates, Inc. were coming to talk to all the employees. At the meeting of all 52 employees, the two polygraph men reviewed the situation in the store and said it would be appreciated if all would take the tests. They said information given during this test would not cause anyone to lose his job nor would anyone be prosecuted.

"Mack came home for the first time in months with a light heart and stated, 'Well, thank the Lord, for once, people will truly know me for what I am!' He was so confident his test would be good because he knew he had nothing to hide. For people to finally know for a fact how clean he lived made him feel good inside. Tuesday was his day off, and he was anxious to get to work on Wednesday and take his tests. But then he learned that everyone who had taken the test Tuesday was jumping up and down. No one agreed with their test and Mack began to get concerned, but he still knew he had nothing to hide so he tried not to worry. I went to pick up my husband for lunch on Wednesday (we only have one car, a 1968 Buick), and it was after noon when he was released from the motel room where the test was given. He came out holding his chest, and I immediately knew something was wrong. I called to him and he just leaned against the car for awhile, got in and stared into space. After what seemed to me like an eternity, he said, 'I have to go back to the store and clock out for lunch.' That's all he said, just stared as if in shock. He's a quiet person anyway and I could sense he was too full to talk. About half-way home (we live in the country on my father's farm) he said, 'There just *ain't* no way—there just ain't no way! Honey, tell me this is just a bad dream and I'm going to wake up in a few minutes.' I asked if he felt like talking about it—he just stared (I was driving). When we got home, he went straight to the bedroom, threw himself on the bed and just lay there, no motion at all. I let our little girl, Dawn, age 4, go out and play—our three boys were in school—so it might be easier for him to find the right words to begin. Finally he began to tell me as best as he could remember.

"The polygraph men had told them all Monday that they were supposed to decide over night how much of the losses each one might be responsible for himself over the past six months. Mack had decided $100. How he came up with that figure is when bags of merchandise such as grits, sugar, rice, detergent, dog food, flour or meal is damaged in shipping or in the stock room, instead of just throwing the good part away, Mack will bring the balance home for me to use. Everyone in the store knows this and he divides any of these things with his men if any of them want it. But the man asked if he was responsible in *any* form that might have cost the store, and this is what he felt he might owe, even though these items were charged off to damaged. The first test that the man gave him went like this.

QUESTION 1. *Are you ready for this test to begin?*
      Mack answered, 'No.' (This is because on Monday night at the meeting the polygraph people said everyone would have to answer 'No' on this question.)

QUESTION 2. *Are you completely convinced I will only ask you the questions I read to you?*
      Answer, 'Yes.'

QUESTION 3. *Other than your job, is there anything else you are afraid I will ask you a question about?*
      Answer, 'No.' (Mack told me—'why should I care if he knew anything about me? I had never laid eyes on him until this day.' All people, sometime in their life, if it's nothing but jokingly making a comment about some sexy model or something, maybe wouldn't want his wife or his mother to know about it, but certainly he could care less if this man knew.)

QUESTION 4. *Between the ages of 10 and 30 do you remember stealing anything from someone who trusted you?*
      Answer, 'Yes.' (Mack said he couldn't remember anything right offhand, but while in school he traveled with the baseball team—his school was State Champions for three years in a row—and he said, 'You know how a bunch of boys are away from home; we'd do anything we thought was funny,' and it was possible during all this time he had picked up something or was responsible for destroying something of value to someone—so to be sure to give a 100% truthful answer he said 'Yes.' He was the team's first baser for East Clarendon High in Tubeville, S.C. and graduated the same year I did in Lake City, 1960.)

QUESTION 5. *Are you sure that $100 will pay Piggly Wiggly for all of the things I asked you about?*
      Answer, 'Yes.' (Mack said his responsibilities cover so much of the store that he said $100 just to be sure everything would be covered.) Then the polygraph man tore the paper from the machine and he said, 'That figure of $100 that you gave me is incorrect.' Mack said he got upset and asked, 'Well, well if $100 isn't right, can you find out what is

and tell me?' The man told Mack to pick another figure. Mack said, 'I have no idea. I was sure $100 would more than cover it. For *sure* in only six months, no more than $200.' So then he did another test.

QUESTION 1. *Do you believe $3,000 will cover what you have contributed as shortages to this store?*
    Answer, 'Yes.' (Naturally this figure would cover Mack's part and lots of other people's part too, as far as he was concerned.)

QUESTION 2. *The same question, but $6,000.*
    Answer, 'Yes.' (Same reason.)

"Then Mack said they asked him all sorts of figures like $50,000, $125,000, and he is almost certain they got as high as $175,000. I told Mack, 'My Lord, Honey, why didn't you think to tell him to ask you one million dollars?' He just said he was so shocked he doubted he couldn't even remember his name. Mack said then he tore the paper from the machine and shook his head and he said, 'There is something bad wrong here, *bad* wrong! Let's take another test.' Mack said okay. This test was like the first one, but this time the last question was, 'Are you working in collusion with anyone to steal from Piggly Wiggly?' Mack said the man explained he meant was Mack in with vendors and would sign for merchandise not delivered to the store. Mack said of all the questions he was asked before, he was more sure of the truthfulness, as 100%, of this question than all the rest because some could have been answered as 'maybe' instead of 'yes' or 'no.' He said the man told him, 'The needle went *wild* on that one!'
"And then after all that, the polygraph operator said, 'Son, if I were you, I'd go back around to that store, sit down and *confess* to Mr. Lewis about the things you've done—and maybe, just maybe, you and Mr. Lewis can work things out and help you keep your job.' When he went back to the store that afternoon, Mack sat down and told Mr. Lewis all he could remember about how he did and what the man said. On Thursday after school, a boy that works part-time in the store came to my husband and told him, 'Today, little Billy Lewis (he's 17) sat up in the classroom and said Mack Coker had stolen $23,000 worth of stuff.' Another boy, Mark, was in that class and Mack called Mark to the office and said, 'Mark, you don't have to say anything except "yes" or "no." Did little Billy say that in class today?' and Mark said 'yes, he did,' and when someone said where in the world is he hiding all that stuff, little Billy replied, 'Oh, he has a brother-in-law that owns a store in Myrtle Beach.' Later that same day, two female part-time employees attending our only private school, Susan and Marian, went to Mack and told him that Connie (Mr. Lewis's only daughter, 18 and a senior in the private academy) told stories something to the same effect. By the time the students got home, can you believe my mother's phone was constantly ringing? (We don't have a phone ourselves, just go next

door and use Mama's.) Some wanting to help, some wanting to find
out what in the world has caused Mack to do something like this?

"The next two days, Mack continued to go to work as usual and
talked to no one but me. He was in such a state—I'd wake up in the
middle of the night to find him sitting on the steps or standing in the
yard. I prayed, 'Dear Father, please don't let me wake up one night
with no life in his body!' On Friday, Mr. Lewis came to him and said,
'Mack, get a cup of coffee, we've got to talk!' He told Mack that Mr.
Gober (he's Vice President of Piggly Wiggly in Charleston) said that
Mack was *solely* responsible for the mess in that store. Mr. Gober said,
'These tests are *right*, because every time we've had trouble in any of
our stores and sent the polygraph in there and fired the people, that
store always straightens up.'

"But I think when I really 'hit bottom' was after Mack was let go
and we went to see this lawyer and he said without a written contract
an employer can just fire you out of the clear, blue sky for your looks,
or because you had your hair parted wrong. He informed me that
polygraph could not even be mentioned in the courtroom concerning a
case. Mack even signed a paper permitting them to give a test, he was
so sure it was the thing to do at the time, and he truly had nothing to
hide. But since it turned out the way it had—why didn't he have the
right to prove it as non-truth and clear his name? I thought everyone is
innocent until proven guilty? Dear God, help us all, if this is the justice
in law we have to live by!"

REPORT OF A POLYGRAPH EXAMINATION, H. ROACH
EXAMINER FOR TRUTH ASSOCIATES, INC.
TO MR. JOHN GOBER, PIGGLY WIGGLY CAROLINA,
STATEMENT OF VOLUNTARY CONSENT TO A POLYGRAPH EXAMINATION

I, *Mack Coker*, do hereby, voluntarily, without duress, coercion, promise
of reward or immunity, agree to a polygraph examination in reference
to *Job Honesty*. I have been advised of my rights under the Fifth
Amendment to the Constitution of the United States. I further authorize the placement of the polygraph attachments to my person in order
that the proper recordings may be made on the polygraph chart. I
hereby release Truth Associates, Inc. and *Piggly Wiggly #48* from any
and all claims or causes of action of any kind that may result from, or
arise out of this polygraph examination. I authorize the release of the
results of said examination to *Mr. John Gober, Piggly Wiggly Carolina*. I
further agree that the results of this Polygraph Examination may be
used as evidence in a Court of Law if the need arises.

WITNESSED BY *H. Roach*, Examiner   SIGNATURE *Mack Coker*

EXAMINATION RESULTS: Prior to the Polygraph Examination, COKER admitted that his theft of money and merchandise from the store since July, 1974 amounted to $100. The results of the initial Polygraph Examination reflect that he was untruthful in the estimate of $100. Additional tests reflect the actual sum to be between $8,000 and $10,000. One of the questions asked COKER during the Polygraph Examination was, "Are you working in collusion with anyone to steal money or merchandise from the Piggly Wiggly?" Answer: NO. The polygrams indicate that he was untruthful in his answer to this question.

PAID ADVERTISEMENT ENTERED IN THE LAKE CITY CLARION FOR MARCH 18, 1977.

A slander suit brought by Douglas M. (Mack) Coker involving the Piggly Wiggly Store in Lake City, South Carolina, was settled out of court recently for an undisclosed sum. The suit arose because of certain defamatory statements about Coker that were circulated around the community some two years ago. In settling the case, the defendants admitted that various rumors had circulated throughout the community about Coker and that from an investigation into this matter, it was determined that these rumors were absolutely false and had no factual basis. A spokesman for the Piggly Wiggly Store said he did not understand how the stories could have spread over Lake City, but that the store is glad for Mack Coker that the matter has been cleared up.

I could cite a great many additional examples much like this one, most of which did not have such happy endings (Mack Coker got his job back). An Indiana minister wrote me about one of his parishioners, Wayne K., a religious young man who had continued to tithe to his church while putting himself through college. Now a graduate accountant, a family man, and coach of the church baseball team, Wayne had a record of six years' loyal service in the local bank. When $2,000 turned up missing from the bank vault, a private detective agency urged that all employees be given lie detector tests. Wayne was the third person tested; he failed the test and he was fired. Community protest forced the bank to have him retested by the Indiana State Police. This time he passed and was reinstated. But on the principle of best two out of three, the bank sent Wayne to the Reid firm in Chicago, where he flunked again and again was fired. His pastor asked me what Wayne might do to prove his innocence. His condition was desperate. After more than a year, each time he thought he had found a new position in the only field he knew, word came from his previous employer and the door slammed in his face. In this instance, I cannot prove that Wayne did not suddenly set aside the habits of a pious lifetime and decide to heist the bank vault, but let us remember first principles: *Neither Wayne nor his pastor nor I have to prove that he is innocent.* There has been no indictment here, no trial before a jury of his peers, not one shred of legally admissible evidence against him—in short, no due process as prescribed by the Fifth, Sixth, and Fourteenth Amendments to the Constitution. And yet, Wayne K. was summarily punished, punished indeed more severely than the law typically punishes the first-time felon who has been properly tried and convicted in criminal court. He has lost his job, been shamed in his community, and been effectively blacklisted from the profession for which he had trained himself in college and in six years at the bank.

Linda K. worked as a bookkeeper for the Kresge Company in Michigan. A polygrapher was called in to investigate losses and he summoned Ms. K. to his motel room for testing. After the lie test was completed, first the polygrapher and later Kresge's security officer put the pressure on, telling Linda that the polygraph showed she was guilty, insisting that she sign a confession and make restitution. Ms. K. refused, denying any theft from Kresge's, and that was her last day in their employment. Feeling branded by the voodoo of the lie test, Linda was afraid at first to tell her mother what had happened. For two weeks she left the house each morning as if going to work and then spent the day weeping at the home of her married sister. Like Wayne K., Linda K. discovered that having lost one job after a trial by polygraph makes it very difficult to find another. After nearly two years, she finally found an employer who did not believe in the lie detector and would hire her as his receptionist. (She also found an attorney who sued Kresge for "reckless infliction of emotional distress" and for "malicious libel." In November of 1979, the jury returned a judgment against Kresge for $100,000 in damages.)

What happened to Mack Coker, Wayne K., and Linda K. was wrong, un-American in the literal sense of that much-abused term, whether the lie tests were invalid (as I believe) or not. If the reader finds himself musing, "Well, maybe they did take the money, in which case they got what was coming to them," consider the case of John K. John is a deputy sheriff in Minnesota, who, in 1975, was serving as a jailer at the county jail. A federal prisoner, a bank robber, was being boarded at the jail while serving as an informant in another investigation. One day he handed over to the federal marshal a loaded gun and a hunting knife, asserting that he had bribed a jailer to smuggle them in, but had changed his mind about escaping. He named John K. as the arms smuggler and offered to prove his story on the polygraph. The sheriff asked Deputy K. if he also would submit to a lie test and, of course, he agreed. Both men were tested at the Minnesota Bureau of Criminal Apprehension; the bank robber passed his test with cool aplomb but John K. flunked his.

Fortunately for John, the sheriff had received a letter just a few weeks earlier from a local merchant commending this deputy. John had been shopping for a present for his young son, and the gift was accidentally wrapped in a paper bag in which another clerk had placed the cash register receipts, several hundred dollars in cash, and checks. Discovering the error when he got home, John called the store and explained, offering to drop off the money on his way to work. "I always thought these lie tests were just about infallible," the sheriff told me, "but why would a man like that accept a bribe to smuggle weapons into the jail?" The bank robber, meanwhile, had been moved to another jail and had escaped from that less secure facility, suggesting a motive for his charge against John. I told the sheriff that the lie detector itself could lie, and often did, and he proceeded with a conventional investigation. Through an informant, he learned that another prisoner, a young man on a work-release program who spent nights and weekends actually in jail, had been running a smuggling operation. That young man confessed that he had brought in the weapons. Ultimately, the bank robber was apprehended and he also confessed that he had fabricated the charge against Deputy K. *Both* lie tests were wrong.

An interesting feature distinguishes those earlier examples in which the polygrapher moved in like Sherlock Holmes and solved the crime with certainty and dispatch, from these other, Kafkaesque trials-by-polygraph in which citizens have been cut off from their employment, branded as deceptive. In the earlier examples, the polygrapher elicited a confession from the guilty person. In the horror stories, the only "evidence" against the victim was the opinion of the polygrapher. We are dealing here with specific-issue situations where the examiners used either the "human lie detector" approach or the Control Question Test, which, as we have seen, is strongly biased against the innocent respondent even when the tests are conducted and the charts evaluated by experienced examiners with Ph.D. degrees.

A Question of Rights

If my employee steals from me or deliberately damages my property, no one will dispute my right to discharge him whether or not I also bring a criminal complaint against him. Unless some contractual agreement between us intervenes, I may also fire my employee merely because I suspect that he might have been guilty of such an action, or, indeed, I may fire him for no reason at all. Apart from humanitarian considerations, there are two factors that work to prevent me from capriciously using this power that I have over his job security. First, it is inconvenient and expensive for me to replace an experienced worker. Second, such arbitrary behavior would damage my reputation as an employer and lower the morale of the rest of my work force. Therefore, I will not fire him unless I think he has given me cause, and I will prefer that my action will also seem justified to others.

Enter the private polygrapher. The employee's confession of theft or vandalism will satisfy both of my requirements. But suppose he does not confess but merely "fails" a lie test? That is likely to satisfy my requirements also. Put yourself in the position of my former neighbor, the president of a large Minneapolis advertising agency, when an $8,000 movie camera used in filming TV commercials was discovered missing from a locked cabinet. To avoid adverse publicity—and to have some hope of prompt action—he called in the private detectives and they brought along their polygraphs. Three employees who had keys to the camera cabinet were given polygraph examinations. Two of them passed but one, Walter K., was diagnosed as deceptive. Walter K. was a talented young film producer, the highest-ranking black employee in the agency, and my neighbor hated to think of losing him. "I wish I'd just written off the camera. Now I'll have to fire Walter and it will be the end of him in this business." In this instance, the polygraph itself undid the mischief it had almost caused. Before any action was taken against Walter, a fourth person was located who had access to the camera. In the course of his lie test this man confessed that he had taken the camera, which was later returned. Walter was exonerated. Except for this lucky circumstance, Walter would have lost his job and perhaps his career. This particular employer was reluctant to take that step, but he would have felt not only justified but obligated to take it, and his other employees would have endorsed it. Walter had, after all, been found guilty—by the mysterious machine.

What if the real thief had not confessed and Walter had been fired, applied for work at another agency, and that agency had called my neighbor for a reference? "A camera was stolen and Walter flunked the lie test so we had to let him go." That would guarantee to blacklist him, would it not? On the other hand, believing him to be a thief, would my neighbor have the right to hide that "knowledge" from his colleagues in the industry?

Let us now consider Walter's rights. He could have refused to take the polygraph test in the first place, refused to sign that "voluntary" agreement form permitting the polygraph company to report the results to his employer and waiving all rights of redress. But this is a hollow right since, under the circumstances, refusal to submit to the lie test must inevitably appear to Walter's colleagues and his employer almost as an admission of guilt. The circumstances that thus constrained Walter's decision to submit were these: (1) the polygrapher represented the lie test as being nearly infallible; (2) this representation was accepted by Walter's employer and colleagues; and (3) Walter was innocent of wrongdoing and therefore expected that he had nothing to fear from submitting to the lie test. This acceptance by all parties of the validity of the lie test is the central issue. If Walter had been asked to stake his career on a role of the dice or the verdict of a Ouija board, his common sense would have informed him to refuse, nor would his refusal have condemned him in the eyes of others.

What if the examiner had actually used a Ouija board, gone through the forms of a standard polygraph test, but based his diagnosis on magic or chance? More specifically, suppose that the polygrapher reported to Walter's boss that Walter's denial of having stolen the camera was deceptive according to the criteria of an accepted lie detection technique of proven high validity, when in fact, the diagnosis had only chance validity. Under these circumstances, the polygrapher's report would appear to constitute defamation and the examiner would be liable for damages. If the employer is negligent and engages an irresponsible or incompetent examiner, then the employer might himself be liable.

But polygraph examiners do not use Ouija boards. They use instead a procedure—either the clinical lie test or the polygraphic lie test—that may be slightly more valid than chance but not nearly so accurate as the 99%, 95%, or 90% that is typically claimed. Suppose that Walter K. had been granted the right of *informed consent* and that the release form he was required to sign had contained a paragraph like the following:

> I have been informed that, according to the best available evidence, the polygraph test I am about to take has an average accuracy of from 63% to 71% and that, if I am truthful, the probability of my being diagnosed as deceptive may be as high as 50%.

It seems safe to say that most people would not submit to testing if they were provided with this information. Suppose, alternatively, that Walter had been encouraged (as he was) to believe that the lie test is "almost infallible," but that the polygrapher had then reported his diagnosis to the employer together with an acknowledgment that about one out of three deceptive diagnoses are erroneous—false-positives. This might protect Walter in some measure but it would have the disadvantage of ruining the polygrapher's business; what employer would pay $300 (and up) for such ambiguous information?

Thus, a question of rights seems to turn on the issue of validity. If you confess that you stole the camera, or if the polygrapher actually sees you take it, then he does not violate your rights by reporting your guilt because his information is valid. Should he report that you are guilty or deceptive without any evidence at all, then he has defamed you and is vulnerable to civil suit. Somewhere between these two extremes, not yet clearly defined in law, there is a crossing point, a degree of validity that would justify a reasonable and prudent man in making such an adverse report about you. If the polygrapher makes such a report, and, either mendaciously or through ignorance of the relevant research, if he indicates that his report is based on an extremely valid test when the best evidence shows only marginal validity, then that misrepresentation might itself be considered a cause for damages.

There have been surprisingly few lawsuits brought on behalf of plaintiffs who had been subjected to trial by polygraph, adjudicated deceptive by the machine or the examiner, and then punished by their employers. Many attorneys, like the first one whom the Cokers approached, seem to assume that there is no cause of action in such cases. Without a contract, the employer can fire whom he wants to. The polygraph tests were, after all, taken "voluntarily." Many victims of such lie test miscarriages, like Wayne K., feel dishonored and on the defensive and assume that they must somehow prove their innocence or take the consequences. If more persons who have been victimized in this fashion were to bring suit against their employers and the polygraphers involved and were successful in obtaining damages, then this particular misuse of the lie test might become less frequent.

To Catch a Thief

But what about all those confessions, those cases of internal crime solved expeditiously by polygraphic interrogation leading to confession by the culprit? If we are to protect the innocent from victimization through trial by polygraph, shall we have to do so at the cost of prohibiting private polygraphic investigation of this type of crime? That was my conviction when the first edition of this book was published, and I had expected it to be the decision also of the U.S. Congress when it passed the Employee Polygraph Protection Act seven years later. I expected the Congress to place employee crime, like other categories of crime, in the hands of the official police. Unhappily, however, the 1988 statute specifically exempts cases in which "the test is administered in connection with an ongoing investigation involving economic loss or injury to the employer's business, such as theft, embezzlement, misappropriation, or an act of unlawful industrial espionage or sabotage."[3] Thus, none of the scandalous abuses reviewed above would have been prohibited under current law.

We are talking here about crime and punishment, deep matters that liberals and conservatives alike have regarded as responsibilities of the State. "To establish justice and insure domestic tranquility" are among the reasons why men establish governments in the first instance and, in the Preamble to the American Constitution, are even given priority over such objectives as providing for the common defense and promoting the general welfare. Private vendettas, vigilante justice, lynch mobs, and hired guns are universally regarded as uncivilized, as symptoms of a society out of control. A business trying to protect itself from internal thievery or sabotage is hardly a "lynch mob," nor is a private polygrapher a "hired gun." But it is manifestly dangerous to delegate the police powers of the State to private agencies.

This line of argument assumes that the innocent employee's rights will be better protected if internal crimes are turned over by his employer to the official police. It assumes that the police will not communicate unproven suspicions or inadmissible evidentiary details to the employer, but will investigate each matter in an orderly fashion leading to the detection and prosecution of the guilty. The police may use polygraphy or may elicit confessions by skillful direct interrogation. But their responsibility is to society generally rather than to the employer in particular, and there are established—if not always effective—methods of monitoring the behavior of the police and curbing abuses of police authority. If anyone doubts society's ability to control the police, these doubts make a poor argument for wider delegation of police power to private and still less accountable hands.

The present apparent inability of the official police to cope with the internal crime problems of American business appears to stem mainly from economic considerations. It seems probable that those businesses with the greatest problems do not contribute enough in local taxes to sustain the expanded police facilities that this plan would require. But tax rates can be modified to meet the need. A manufacturer whose product is being sabotaged, a business that experiences the theft of an expensive piece of equipment, a retail merchant whose records indicate recurrent losses of internal origin, all of these, in a well-regulated society, should be able to obtain prompt and effective police service. Much has been said, by advocates of private polygraphy, about the rights of business to protect its plant and property. I agree with these claims, but I would argue that business, like the ordinary citizen, should turn to the officially constituted police agencies for the protection of those rights rather than taking the law into its own hands.

Chapter 15

PREEMPLOYMENT SCREENING BY THE FBI AND OTHER FEDERAL AGENCIES

The whole process smacks of 20th century witchcraft.... The burden of proof should be on those who assert the efficacy of polygraph in predicting the behavior of prospective ... employees. There have been practically no efforts to compile this proof.... Why then do [employers] have such blind faith in these devices? In my opinion, it is directly related to the role of science and technology in our society—the cult of the "expert." There is an increasing belief that anything scientific must be more reliable and rational than the judgment of men.... There is no necessity for these infringements of freedom and invasions of privacy; but even if there were a necessity for them, I believe that every citizen should answer like William Pitt: "Necessity is the plea for every infringement of human liberty. It is the argument of tyrants; it is the creed of slaves."

—SENATOR SAM J. ERVIN, JR.

The federal Employee Polygraph Protection Act of 1988 eliminated most preemployment polygraph screening by American business, but drug companies, private security businesses, and federal, state, and local governments were exempted from the Act's proscriptions. For many years, the CIA and other federal security agencies have required their employees to submit to polygraph screening, both prior to hiring and periodically afterwards. Because of the polygraph's effectiveness as a "bloodless third degree" (see Chapter 18), it is likely that evidence of significant misconduct was uncovered from time to time through admissions made under the stress of these examinations. In its 1983 report, "The Accuracy and Utility of Polygraph Testing," the Defense Department stressed the utility of the lie detector as an interrogation tool. Claiming only that the polygraph's accuracy is "significantly above chance," the report lists numerous examples of damaging admissions elicited by polygraph examiners. One applicant for a job with the supersecret National Security Agency, for example, "admitted that his engineering degree was phony (he bought it through mail order from London for $100)," that he "shot and wounded his second wife," and that his present wife "is missing under unusual circumstances that he would not explain."[1]

But it is only prudent to suppose that actual spies sent to or recruited in the United States will have been trained in how to cope with polygraphs: "Never admit anything significant; augment your reactions to control questions by self-stimulation" (see Chapter 19). Aldrich Ames and Harold Nicholson apparently passed CIA polygraph tests so convincingly that their superiors turned a blind eye to, for example, Ames's unexplained wealth, his luxurious home and fancy car. It is reminiscent of the story of the Emperor's New Clothes—once you have passed the polygraph, your associates can no longer see the naked truth.

The Case of Major C

But my concern in this chapter is with the other side of this screening coin: Who is it that we are screening out of our police and intelligence agencies with what Senator Ervin called "20th century witchcraft"? I was alerted to this problem by a curious phone call I received in the spring of 1983. The caller identified himself as Major C, a U.S. Army officer stationed in Washington, D.C. He was calling to ask whether, if he flew to Minneapolis, I would be willing to meet with him to discuss "Chapter 19 in your book." Major C was a West Point graduate who had specialized in Soviet area studies. He spoke Russian fluently and had served as a translator on the "hot line" between Washington and Moscow. With officer colleagues under U.S. Army Russian Institute (USARI) auspices, he had spent time in the Soviet Union, meeting Russian citizens and improving his knowledge of the language and the culture. With another officer, a Major X, he became acquainted with some Russian dissidents in Moscow who later managed to emigrate to the United States where they contacted Major X. This contact Major X duly reported to Army Intelligence. This led in turn to both officers, because they held high SCI (Sensitive Compartmented Information) security clearances, being called in for polygraph screening. Major X passed his interrogation without problems, but Major C was found to be deceitful in denying contacts with foreign intelligence agencies or divulging secrets to a foreign national.

Major C's security clearances were immediately revoked and he was told to "go home tonight and make a list of everything you've done or thought of doing over the past two years that you might feel guilty about. Think of it as an act of mental cleansing. Bring that list with you tomorrow and we'll go over it, get it all out on the table, and then perhaps the polygraph will give you a clean record." But Major C was a true West Point, Honor Code, straight arrow, and his list of sins was short. For example, he recalled a recent day when the *Washington Post* had reported on a hot line communication with Moscow. Major C dined that evening with Major X and his wife, who asked Major C if he had handled that communication and he answered "Yes." Mrs. X had been born in South Africa and was, arguably, a "foreign national." Even after "cleansing his mind" of these misdeeds, he still responded on the polygraph to questions about foreign nationals and intelligence agencies.

The reason Major C proposed to come consult with me was desperation. He had served honorably for 16 years, had been as honest as he knew how with the polygraph examiner, and yet he'd been betrayed, the polygraph apparently had lied about him. He had found the first edition of this book in the library ("know thy enemy") and now was determined to fight back, to try to save his career. This machine, or the person running it, had cheated him, so now, as a last resort, he was prepared to try to cheat the machine. I told Major C on the phone that a trip to Minneapolis would be a waste of time and money, that I had nothing useful to add to what was in the book. I did, however, agree to send him an affidavit explaining why, as an expert on these matters, I believed that his polygraph results had no probative value.

Some weeks later, I got another call from Major C and he sounded like a different person. His Top Secret clearance had been restored, although not the even higher SCI clearance. His promotion to Lt. Colonel had come through. It had been decided that his years of training as a Soviet area specialist would be set aside, and he regretted that (so did I since my tax money had helped defray that million-dollar investment). But he was to be given instead a year-long course in Turkish and then sent to a station in Ankara and he was looking forward to that. Major C subsequently provided me with a copy of his file on this debacle, from which it seems clear that the only "evidence" against him, from the beginning, was the vague suspicion expressed by polygraph examiners.

One illuminating example of Major C's character was an incident he recounted from that USARI trip to Moscow. One night, with his companion officers, all dressed as civilians, he returned to the hotel from dinner at a restaurant. His colleagues knew that the night-shift Russian security agent would be on duty in the hotel lobby and that he would ask for their identification. The other officers thought it might be fun to pretend that they were Russian visitors from out of town, just to see if they could get away with the charade. The idea made Major C uncomfortable and he insisted on going in separately without any pretense. This is the kind of straight arrow that I believe is especially vulnerable to betrayal by the polygraph. It is also the kind of man I would like to see employed by my government in sensitive positions, men and women whose lively consciences will prevent them from abusing their authority.

Cry Havoc

While we know that Richard Nixon valued the lie detector as a tool of intimidation, it was during Ronald Reagan's incumbency in the 1980s that the president decided to "cry havoc and unleash the dogs of war," that is, to sic the FBI's polygraphers on administration officials suspected of leaking information to the press. In 1982, for example, George Wilson, a veteran reporter for the *Washington Post*, revealed that the Defense Resources Board, a group of 30 high-level Pentagon officials that managed the defense budget, had secretly concluded that the Pentagon would have to ask Congress for $2.25 trillion over the next five years to accomplish what Reagan had said would be done for "only" $1.5 trillion. The outraged president decreed that most of the board members, including Navy Secretary John Lehman and the chairman of the Joint Chiefs, Gen. David C. Jones, would have to submit to polygraph "fluttering." It is interesting to imagine oneself in the place of the FBI polygrapher given this assignment.

Figure 15.1. Even honest people cannot always relax when asked accusatory questions while hooked up to a polygraph. © Gahan Wilson, used by permission.

Nearly every high official tested would be likely to make the polygraph pens do a dance when asked that question that could write finis to their careers. No examiner in his right mind would be likely to identify the alleged culprit until all had been tested. Then he could compare the charts, looking for the one that indicated the strongest reaction, and hoping that it would be a lower-level official than the Navy secretary or the chairman of the Joint Chiefs. All we know for sure is that the designated leaker turned out to be John Tillson, director of Manpower Management at the Pentagon. Tillson ended up flunking three separate polygraph tests, at which point he got hold of a copy of the first edition of this book and called me for advice. There was not much I could do, of course, but the journalist, Wilson, saved the day (and Tillson's job) by taking the unusual step of writing to Secretary of Defense Caspar Weinberger. "An honorable man stands falsely accused," he wrote. "… I give you my word, John was in no way connected with the story I gathered and wrote."

So strong is Washington's faith in the myth of the lie detector, however, that events like the Tillson debacle are brushed off as aberrations.

After all, most people whose careers are smashed by the power of the polygraph do not have the good fortune to be able to prove that their test results were wrong. Toward the end of 1982, a Marine colonel, Robert McFarlane, failed a lie detector test seeking the source of a leak to the *New York Times* about a British spy scandal known to the American, British, and Soviet governments but which our National Security Council, for which McFarlane worked, wanted to conceal from the public. McFarlane managed to persuade the *Times* publisher, himself a former Marine, to assure Reagan by telephone that McFarlane was not the source. The highest-ranking official to be victimized (so far) was Michael Pillsbury, fired from his job as assistant undersecretary of defense in 1986 because he failed a polygraph test relating to the leak of a plan to sell Stinger missiles to the rebels in Afghanistan. In his case, too, a journalist revealed that it was two senators who had been the source of the leaked information. Pillsbury's reputation and his security clearances were ultimately restored, but not his job.[2]

Screening FBI Agents

As we have seen, the FBI has used the polygraph for years as an investigative tool, but it was not until March of 1994 that Director Louis Freeh instituted the practice of polygraph screening of all new agents. Inasmuch as the Cold War was over and it was the CIA, which had long used polygraph screening, that was experiencing the problem of moles boring from within, it is difficult to understand the director's mind on this. Soon after, my telephone began to ring with calls from young people all around the country who had dreamed of working for the FBI only to be turned away at the last moment because the polygraph pens did a suspicious dance during their preemployment screening test. Almost invariably it was the questions about illegal drug use that caused the problem.

In April of 1995, I received a letter from a young woman in California. It began as follows:

> I am writing because I need your help.... First, you must know that I prepared myself most of my life to be an FBI Agent. I come from a law enforcement family. After graduation from Law School, and then from Graduate Business School, I applied to the FBI. After eight months in the hiring process, during which I passed all the separate steps and

came out on top, I had to take a polygraph exam. I failed, but I was telling the truth! I could not believe it—I was so shocked.

Elizabeth M. enclosed with her letter a six-page, typewritten memoir she had composed for her journal after returning from her devastating encounter with the polygraph. She had arrived at the FBI office in the San Francisco Federal Building at 12:15 P.M. where she was welcomed by a Ms. McDermott, who had her fill out some forms and provide a urine sample. Then Agent Feldman tested her to see how many times she could fire a 9-mm handgun in 30 seconds. Next, she was handed over to an Agent Otstott, who took her, first, to be fingerprinted, and then to the polygraph examination room. En route, he told her that one-third of the candidates failed the polygraph and they tended to fail on the drug use question.

> I smiled and shook my head at that, because I knew that would not be a problem for me at all. Still, it made me very anxious to hear how many people failed the test. This was the 3rd time I had been told that statistic by a FBI recruiter and I was very nervous.

Ron, the polygrapher, seemed friendly at first. He asked her some questions without the polygraph attachments, then with the machine in operation. She was nervous and he commented about problems with her breathing record. After the first set of questions:

> Ron came to sit in front of me again. He said that there had been a reaction on the question of whether another group had sent me to join the FBI (in other words, that I was sent to be a double agent) but then he said that he was not worried about that because I did not seem like a spy to him. He seemed odd then, like he controlled my whole future, like he determined if I was lying, not the machine. I got the feeling either that he did not really like me very much, or that he was patronizing me.

Ron then asked her another set of questions without the polygraph; these included questions about drug use. She had to "check some boxes" on a form and sign her name.

> I just thought that if it had anything to do with drugs, it was a no-brainer—after all, I had never done any, so just sign it…. After all, I was raised strict Roman Catholic in a Law Enforcement family, which emphasized education. Drugs were not just eschewed by my family and myself, but vehemently despised. In fact, just the thought of my taking an illegal drug makes me nauseous, sick to my stomach. It's an automatic reaction that I have had since a very young child when my

father would come home from work and tell us stories of children and young people killed and destroyed by drugs.... I never experimented or took drugs.... When on occasion I went to a party where drugs were present, I would feel uncomfortable and leave—I just could not stand to be around them.

Ron obviously found it hard to believe that she had never even tried marijuana.

"Most people in your generation tried drugs, we understand, its OK to admit that." He persisted in acting incredulous when I stuck by my statement of no drug taking; finally he sighed and hooked me up to the machine again.... I was upset and angry that he did not seem to believe me.... He just acted like I was a liar before he even hooked me up. I answered the next series of questions which dealt with drugs, driving while drunk, stealing, and other similar questions.... When he asked me whether I took illegal drugs, I was pissed off and sick. But I thought also, see, now I will show him I am telling the truth. But when he was finished he came to sit in front of me again and told me that there was a reaction to ... the drug use question. I was dumbfounded!

We are still only on page four of this six-page account. Ron is just getting into his stride. "He kept saying repeatedly, over and over, 'you have a cancer inside of you. The cancer is growing. We need to cut it out. I am here to cut it out. If you just tell me what you have not told me yet then we will have cut that cancer out of you.'" (We can see here that Ron is of the Robert Ferguson school (Chapter 10) and thinks of himself as a surgeon of the soul.) "'You have a lesion in you. It is growing and I have to lance it to kill it. If you tell me about your drug use, everything will be all right.' This went on for hours and hours—literally. Frustration seeped into every pore over the hours that I was there asserting the truth. Tears fell into my mouth and I tasted salt and sorrow with each moment."

Elizabeth remembered an incident about age 12 or 13 when she tried smoking a cigarette. And another when she was a student in Paris drinking wine with a woman friend who smoked a pipe. She had been a "little tipsy" and had taken a puff on her friend's pipe. None of these innocent recollections satisfied Ron.

As I walked out of the room, I asked, what next? I had been in that room *five hours* being *interrogated* for telling the *truth*. I shook, I was exhausted and sad and angry. I said that I wanted to take the test again, the test was wrong and I want to prove myself. He smirked and said that Washington would make the determination. I now know that he knew what Washington's answer would be. He treated me like a child, cruelly not telling me candidly what the process was once a person failed the polygraph. I lived under tremendous stress and anxiety for three weeks until I learned that the FBI would not allow me to be retested. My application was terminated. Cold, cruel, bureaucratic. Killing an individual's dream, conceived and striven for in innocence and purity of action. Sacrificing an individual's future on the altars of prejudice and arbitrariness.

Elizabeth M.'s journal suggests, on the one hand, that she may be too vulnerable, too innocent to be successful as a professional police person. On the other hand, she was not expecting to be dealing here with Mafiosi or other villains. She had come, full of enthusiastic anticipation, to complete her application to enter the career she had dreamed of from childhood, expecting to be dealing with FBI professionals, representative of her future colleagues, expecting to be treated honorably and fairly. *But*, most important of all, this polygraph screening test is not designed, not intended, and not regarded as a stress test to see which applicants are best qualified to tolerate abuse without losing their composure. This polygraph test is intended and regarded by Director Freeh and his minions as a test of *honesty* and of *veracity*, no more, no less. And no sentient person, with the possible exception of Ron himself and perhaps Robert Ferguson, could read Elizabeth M.'s journal without concluding that it was Ron and his polygraph that did the lying here, not Elizabeth.

During the same month that Elizabeth underwent her trial-by-polygraph, James K. Murphy, the chief of the FBI's Polygraph Unit, submitted an affidavit to a federal court in New Mexico acknowledging that the Department of Justice and the FBI "oppose any attempt to enter the results of polygraph examinations into evidence at trial."[3] As Elizabeth's memoir graphically portrays, these preemployment screening examinations are trials indeed, in which the polygrapher sits as judge, jury, and executioner combined. I wrote to FBI Director Freeh on April 19, 1995, enclosing Elizabeth M.'s journal of her experience together my own curriculum vitae documenting my standing as a scientist and as perhaps the leading scientific critic of this pseudo-scientific technology. I urged him to do a background check on Elizabeth M. to verify her own account of her experience, or lack of experience, with illegal drugs. In May of 1996, I wrote again to Director Freeh, this time on behalf of a young man, an intelligence analyst for the Drug Enforcement Administration, who had been required throughout his tenure there to take periodic urinalyses for drugs, but who also had failed the drug questions when polygraphed in connection with his application to transfer to the FBI. Director Freeh never acknowledged receipt of either letter. I suggest you think about that, dear Reader. Perhaps the director of the FBI cannot personally answer every letter that might cross his desk. But this employee of yours and mine *ought* at least to acknowledge a communication from a world-recognized expert on a technique that Freeh himself had ordered to be used in screening applicants to work as agents in our federal police agency. Is this incompetence?—or is it arrogance of power?

I think it is now obvious that polygraph testing has failed to screen out from our intelligence agencies potential traitors and moles. On the contrary, it seems to have served as a shield for such people who, having passed the polygraph, become immune to commonsense suspicions. And it is certainly obvious that polygraph testing has been screening out some of the very kinds of people one would most want to see placed in positions of trust: conscientious people like Major C, Elizabeth M., John Tillson, Michael Pillsbury, and Col. McFarlane—highly socialized people who seem to be especially vulnerable to false-positive polygraph mistakes.

Chapter 16

HONESTY TESTING: AN ENVIRONMENTAL IMPACT ASSESSMENT

After years of catching thieves with the lie-detector, we've perfected a way to catch them with paper and pencil.

—Advertisement for the Reid Report

We all, like sheep, have gone astray.

—HANDEL, the *Messiah*

I first heard of paper-and-pencil "honesty" tests in 1976, at a hearing before a committee of the Minnesota state legislature on a bill to ban polygraph testing of employees. Sister Terressa, a Roman Catholic nun, asked to testify. She told the committee that she had applied for a job at a B. Dalton bookstore which, at that time, was using a questionnaire called the Reid Report, to screen prospective employees. A week or two after completing her application, Sister Terressa called B. Dalton's to inquire. "I'm sorry," she was told, "I'm afraid you got the lowest score on the honesty test that we've ever seen!" Largely because of Sister Terressa's testimony, the Minnesota statute outlawed not only the polygraph but "any other test purporting to test honesty."

The Reid Report was developed by Reid's polygraph testing firm in

Chicago. The scoring key is proprietary and not available to psychologists like myself who might be interested in making an independent assessment of the test's psychometric properties. Nevertheless, the scoring principles of the Reid, the Stanton Survey, the Personnel Security Inventory, and similar devices are not difficult to infer. One group of items invites the respondent to admit to various crimes and misdemeanors ranging from homicide and forgery to stealing from the company or lying to the boss. These items are supplemented by rating scales similar to this:

> Check the nearest total dollar value of all merchandise and property you have taken without proper authorization from previous employers.
> ( ) $5000   ( ) $2500   ( ) $1000   ( ) $500   ( ) $250
> ( ) $100   ( ) $50   ( ) $25   ( ) $10   ( ) None

The rationale behind such questions is, plainly, that someone who has been guilty of prior thefts and other dishonesty is likely to do it again given the chance.

Can all individuals be counted upon to (honestly) respond to such direct questions, even after tests based on this principle become ubiquitous in preemployment selection? And are all those who respond other than "None" on items like the one shown above really dishonest in any reasonable, normative sense? It is interesting to imagine which alternative one's Heavenly Auditor would check, who had kept a complete list over the years of one's phone calls, stamps, pencils, and so on, appropriated for personal use. *Could* a scrupulously honest respondent answer "None"? If a job applicant does admit to really major misconduct in the recent past, that of course might be reasonable and adequate grounds for rejection even without elaborate empirical validation. But can one say *a priori* whether applicants who check "$50" are more or less honest than those who check "None"?

Supplementing these "admissions" items are sets of questions intended to measure "punitiveness" and "attitudes toward theft." The rationale here is that thieves will not advocate strict punishment of persons like themselves and that the dishonest tend to "project" dishonesty on others. For example:

> How many employees take small things from their employers from time to time?
> 95%   80%   60%   40%   20%   5%

It is assumed that a dishonest respondent will make higher estimates than will a desirable employee.

It was now possible to speculate how it might happen that Sister Terressa got such a low score on the Reid Report. An educated woman and a member of a worldly order, she did not have a sufficiently naive, Polly-annish view of humankind to assume that nearly everyone is always honest, and thus get a really "honest" score on the attitude items. (It is interesting that, since purveyors of honesty tests set their cutting scores so as to fail about half of all applicants, their own view of human nature would prevent them, too, from passing their own tests.) With respect to the punitiveness dimension, of course, Sister Terressa was probably undone by Christian charity.

Some support for this speculation was provided by administering the Reid attitude and punitiveness items, along with the Socialization (So) scale of the California Psychological Inventory, to 41 members of cloistered religious orders in Minnesota. For comparison purposes, the same items were also administered to 184 junior college students and 71 inmates of a Minnesota maximum-security prison. The monks and nuns were least punitive and the students most. The So scale, in contrast, showed a sub-stantially higher level of social and moral commitment for the religious than for the felons.[1]

As we have seen, the use of the polygraph for preemployment screen-ing, which had developed into a significant industry beginning in the 1950s, is now largely prohibited in the private sector of the United States by the Employee Polygraph Protection Act of 1988. Many polygraphers used the polygraph as an interrogation tool, asking subjects to explain why certain questions on the screening test might have disturbed them enough to cause a response. In this stressful context, many people will make damaging admissions hoping that, after their conscience has been purged, the polygraph will pronounce them truthful. These admissions, rather than the polygraph results per se, were often used as the basis for an adverse recommendation. One problem with this practice was that these "damaging admissions" frequently involved petty misconduct of a kind that most people could find in their autobiographies if they looked hard enough and it seemed odd that these applicants were being penalized for being truthful. It is at least possible that seriously undersocialized people, in this situation, would not make damaging admissions. Since no ade-quately controlled studies of the accuracy of such polygraph screening have ever been reported, we cannot say whether their predictive validity was positive, zero, or even negative.

Some polygraph firms had already developed paper-and-pencil "honesty" tests, like the Reid Report, for clients unable to afford or unwilling to use polygraph screening and, since 1988, the market for such alternatives has exploded. The use of psychometric screening of job applicants in industry has a long history. This is an area in which the use of tests with very modest predictive validities may be justified, at least from the employer's point of view. Estimates of business's annual loss to employee theft range from $5 billion to as much as $60 billion.[2] If the latter estimate can be believed, reducing the proportion of potential thieves in the work force by even 10% might save American business $6 billion a year.

If, say, one in ten of all applicants is a bad risk, then a test with a validity of only 53% (vs. chance = 50%) would accomplish that 10% reduction. For example, with 1,000 applicants tested, 47 of the 100 thieves and 477 of the 900 honest persons would survive the screening, giving 47/524 ≃ 9% thieves among the survivors. But it would do so at the cost of also excluding 47% of the well-socialized, potentially trustworthy applicants. This is a cost that the employer sees only as a testing cost, but the excluded job applicants may well evaluate it differently. Since most Americans, or their children, are potential job applicants, the general public is entitled to take an interest in this question. Indeed, the public's stake in the decision of American business to employ honesty testing and its potential impact on our economic and social environment has much in common with the public's interest in decisions by industry that threaten to pollute the physical environment.

The modern response to the latter situation is to call for an Environmental Impact Statement (EIS), which lays out the alternatives and forthrightly examines their probable effects on the commonweal. The 1990 report by the U.S. Congress's Office of Technology Assessment (OTA)[3] seems to me to be an excellent example of an EIS on integrity testing (although it lacks a discussion of other alternatives; see below), whereas the 1991 report of a task force appointed by the American Psychological Association (APA)[4] I found disappointing. Before expanding on this contrast, it is appropriate to acknowledge that the problem of employee dishonesty is real and likely to get worse.

The Problem

Although the United States boasts by far the highest rate of homicide of any industrialized nation,[5] the vast majority of U.S. crimes, as in most developed countries, are crimes against property. Although theft is commonly attributed to poverty, exacerbated by the display on television of unattainable luxuries, most poor people are not thieves. The most plausible theory holds that crime is largely the result of a failure of socialization of biologically susceptible youngsters.[6] Some upper- and middle-class parents can and do fail to socialize their offspring, but it is reasonable to suppose that failure to meet this basic parental responsibility is more likely among immature parents who are undersocialized themselves, among single parents struggling to do the job alone, and, because of the importance of fathers both as role models and as agents of control and discipline, especially among single mothers. There are currently some 15 million children, one-fourth of *all* the children under the age of 18 in the United States today, who are being reared by single mothers.[7] One-third of these children were born out of wedlock and many of these mothers were themselves poorly educated and poorly socialized. The proportion of births to unwed mothers in the United States has increased from about 5% in 1960 to more than 30% today.[8]

These discouraging statistics suggest that crime rates are unlikely to diminish in the foreseeable future. Although the preponderance of active criminals are young, male, and unemployed, some undersocialized people do find their way into the work force where they are presumably responsible for much of the employee theft. The desire of employers to find effective methods of coping with the problem of employee misconduct is therefore understandable. One obvious approach is to try to screen job applicants for their proclivities toward theft and other counterproductive behavior.

The Integrity Screening Approach

There are currently some 45 integrity tests on the market. These screening devices are in use by more than 5,000 firms in the United States and are administered to millions of job seekers annually.[9] The most widely marketed tests have item content very similar to the type already outlined. Many were constructed by nonpsychologists with no training in mental measurement; the Phase II Profile, for example, which had been given to more than a million people as of 1983, was constructed by a former police officer, who claims that his device has a "reliability factor" of .97 out of 1.0.[10] Other scales have been constructed with considerable psychometric sophistication. The PDI Employment Inventory,[11] for example, designed to predict "counterproductive behavior" rather than honesty per se with items representing a dozen or more content areas, is a model of theory-guided, empirically keyed test construction.

The APA task force began "with the position that honesty tests should be judged by the same standards as other measures developed and used by psychologists."[12] I believe this postulate reflects a concern that the adverse environmental impacts of honesty testing might also be found for some accepted practices in the field of personnel selection. Nevertheless, the task force concluded that some so-called honesty tests boast no validation evidence at all, whereas, even in the case of tests for which research exists, "test publishers have relied on the cloak of proprietary interests to withhold information concerning the development and scoring of the tests, along with other basic psychometric information."[13]

Many of the tests in this category include the same three types of items as are found in the Reid Report discussed above. As I have already suggested, it is possible that "admissions" items might screen out both the more naive of thieves and the more scrupulous of honest people. The "attitude" items might exclude those who assume others are as under-socialized as themselves as well as honorable people sophisticated enough to realize that not everyone is as well-socialized as they are. And the "punitiveness" items almost certainly screen out others like Sister Terressa and the monks and nuns in the Minnesota sample. In the absence of solid validity studies, replicated by impartial investigators, the environmental impact of the array of "honesty" tests currently marketed simply cannot be assessed. Yet, in respect to these devices, the task force concludes: "We do not believe that there is any sound basis for prohibiting their development *and use*; indeed, to do so would only invite alternative forms of pre-employment screening that would be less open, scientific, and controllable."[14] It does not seem obvious, however, what methods *could be* "less open, scientific, and controllable."

The OTA identified several studies of the predictive validity of honesty tests, all conducted by the test publishers and of variable quality. Let us generously (even unrealistically) suppose that well-designed studies, done by impartial scientists and published in peer-reviewed journals, were to agree in showing that some particular honesty test has a validity in a variety of preemployment screening applications equal to the average found in the five in-house studies reviewed by OTA (see Table 16.1).

On average, these tests failed 56% of applicants, and thereby rejected 79% of the potential thieves, while also rejecting 55% of the honorable job seekers! Only 4% of the job applicants who failed the integrity tests turned

Table 16.1. Totals of OTA's Five In-House
Studies of Integrity Test Validity

	Failed test	Passed test	Totals
Honest employees	3,019	2,471	5,490
Dishonest employees	133	35	168
Total	3,152	2,506	5,658

out to be dishonest. Because of its cost efficiency, however, such a test would be very widely used. A few villains, rejected by Employer A, might make it over the cutting score with Employer B, due to the imperfect reliability of any psychometric device. A few others would, inevitably, learn how to beat the test (which would not be difficult; see below). A few of the many honest people who were "false-positive" errors in their first attempt would also benefit from the test's unreliability, getting a better score later and eventually obtaining a job. But the widespread use of this screening device would inevitably mean that millions of potentially fine employees would be essentially excluded from the work force. The OTA study specifically addresses this issue of systematic misclassification (p. 13) as well as the general question of "What happens to the misclassified?" (pp. 65–66), whereas the task force report does not even consider what to me is the most important public policy concern of all. Without citing actual data like those above, the task force concludes that "the preponderance of the evidence is supportive of [honesty tests'] predictive validity" (p. 26).

The Employment Inventory (EI), published by Personnel Decisions, Inc., and described above as a model of sophisticated test construction, has been validated on thousands of employees in a variety of businesses as a predictor, not of honesty, but of satisfactory employee performance. The development of the EI involved administration of the test to 4,652 applicants for employment in a large chain of retail stores. Without regard to EI scores, 2,988 of these applicants were hired and their employment status monitored for three to six months after hire. Based on careful performance evaluations, the employees were classified into four categories: satisfactory performers (would rehire), marginal performers (have been laid off but might rehire), problem performers (would not rehire), and gross misconduct terminations. Using an empirically derived cutting score, the proportion of each criterion group who would have passed the EI and been recommended for hire is shown in Table 16.2.

The EI is not specifically an honesty test although some of its items are

Table 16.2. Success of One of the Best
Screening Tests for Predicting
Satisfactory Employee Performance

Criterion group	Percent passing
Satisfactory performers	68
Marginal performers	53
Problem performers	37
Gross misconduct terminations	29

similar to those used in published honesty tests. It seems fair to say that the EI represents the state of the psychometrician's art as applied to the problem of predicting employee performance as it is affected by individual differences in attitude, socialization, and personality factors, exclusive of abilities. There is no doubt that the EI is cost effective from the employer's point of view. Yet we must note in the table that 32% of the satisfactory employees and nearly half of the marginal group (at least some of whom did well enough that their supervisors said they would rehire them) would have been rejected on the basis of their EI scores. The test publisher urges that hiring decisions should not be based solely on the test score but that recommendation is unlikely to be honored since it would only serve to decrease the test's cost effectiveness. If the EI's retest reliability is on the order of .80, some good job prospects who failed the test at one job site might pass it at another. But there can be no doubt, should this test or any other similar selection device come to dominate the ever-growing market, millions of our children and grandchildren, honest and potentially good workers, will be permanently excluded from acquiring entry-level jobs. To the extent that the use of *any* single selection device becomes ubiquitous, or any group of similar devices as highly intercorrelated as the most popular "honesty" tests are likely to be, this problem of systematic exclusion will exist.

Beating the "Honesty" Tests

We shall see in Chapter 19 that the conventional lie detector test can rather easily be defeated, both by sophisticated criminals or spies and also by honorable people who understand that merely answering truthfully is no guarantee at all of passing the polygraph. A similar situation prevails with questionnaire "honesty" tests. Since my first acquaintance with the Reid Report, back in the 1970s, I have always thought that just by answering oppositely to the way Sister Terressa honestly answered her test, one should be able to sail through with flying colors. In 1996, a former student of mine, Dr. Scott Lilienfeld, and two colleagues published a systematic investigation of this conjecture.[15] They divided 195 college student subjects into five groups. All were told to imagine that they had graduated and "were applying for a job in which they were very interested." They were to take two questionnaires, the answers to which might determine whether they would be hired. The two tests were an overt honesty test similar in content to the Reid Report and a covert test, the Personnel Reaction Blank (PRB),[16] which is similar in construction to the Employment Inventory, discussed above.

The Just Take group were instructed to answer the questionnaires as they would expect to in real life. The Fake Good group were told to try to "beat the tests" by appearing as honest as possible. Group 3, the Overt Coaching group, were given instruction on how to answer the overt honesty test only. This involved giving them the mnemonic acronym "PLACE," each letter of which stood for a hint about how to answer the test's questions. "P," for "protect," meant that subjects should not appear motivated to protect dishonest coworkers. "L," for "lenient," reminded them that, in their answers, they should not appear ready to go easy on offenders (i.e., they should appear to be as punitive as possible). "A," for "admit," reminded them not to admit any wrongdoing on the test. "C," for "common," meant that they should not answer questions in a way that suggested they believed that even the slightest dishonesty was common (i.e., they should appear to believe that almost everyone was strictly honest). Finally, "E," for "excuses," reminded subjects in the Overt group that their answers should indicate that they believed there could be no valid excuse for dishonesty.

The fourth group, coached in how to defeat the covert test, used the acronym "SCARED." "S," for "safety conscious," meant they should deny interest in risk-taking behavior. "C" meant "conformist," "A" meant "agreeable," "R" was "restrained," "E" stood for "even tempered," and "D" stood for "dependable." These hints reminded the Covert Coaching subjects that their answers were to give the impression that they were strongly endowed with each of these six traits. Finally, Group 5 were coached in how to beat both the covert and the overt types of test.

The results of this exercise were fascinating. The overt honesty test, just as one might expect, was easy to beat. Even subjects told to "fake good" without instruction as to how to do this got significantly better scores than subjects in the Just Take condition. Subjects given coaching in how to beat the overt test elevated their scores substantially in the "honest" direction. The covert test, in contrast, was quite a different story. None of the groups trying to beat that test succeeded in improving their scores to a significant degree.

A total of 12 studies of the susceptibility of honesty tests to faking have appeared and their results summarized in a recent meta-analysis.[17] As expected, overt tests are easily faked even without coaching (although coaching increases the distortion), and in contrast to the findings by Lilienfeld *et al.*, even the covert or personality-based tests were moderately influenced in the "good" direction when subjects attempted to give a good impression. Thus we see an almost perfect parallel with the situation in polygraph testing: A high proportion of honest people will fail lie detector or honesty tests when they rely on their truthful answers to demonstrate their innocence or good character. Meanwhile, the more clever of the villains (as well as those honest people who have read this book!) will employ countermeasures to ensure that they manage to pass.

The Good Management Approach

Anticipating the difficulty of saying anything good about integrity screening, the APA task force prefaces their analysis with the following question: "For any potential problem with honesty tests, one must determine the extent to which alternative procedures used for the same purpose would be similarly indicted. In essence, one must always keep salient the question, 'What would you have them do instead?' "[18] In what seems to have been one of the first published discussions of this problem, I suggested in the 1981 edition of this book that the obvious alternative to integrity screening and its problems is good management. Although I do not claim expertise in this area, the suggestions made then still seem sensible to me.

Security. Every employer can follow the same principle for preventing theft that you and I use when we lock our house or car: He can take steps to minimize the opportunities for stealing. He can keep track of inventory, maintain good records, restrict access to valuables, lock things up, fasten things down. There are two multipump gasoline stations that I frequent, both still using attendants to pump gas. At one station, the amount of each sale is called into the office/store by intercom and the customer pays inside at the cash register. At the other station, each attendant collects and makes change from his pocket. This second arrangement clearly invites theft; the first one makes it considerably harder.

Employee Morale. It would be naive to suppose that good employee relations will eliminate employee misconduct. But one can hardly doubt that it will pay in the long run to treat one's employees with consideration and respect, to pay them adequately and treat them fairly. For the great mass of people, stealing comes hard and requires some sort of rationalization. If management can be regarded as the enemy, selfish, indifferent, unfair, then a ready excuse is provided. One of the best techniques might be some form of profit-sharing arrangement, ideally one that is focused on relatively small groups of employees who are in daily contact, so that each person will be encouraged to feel that to steal from the company is to steal from one's friends and colleagues. The success of South Bend Lathe Corporation in Indiana is instructive. Through the Employee Stock Ownership Plan, sponsored by the U.S. Department of Commerce, the employees and management of South Bend Lathe bought the company in 1974. Earnings per share more than tripled in the next three years—and there is no problem with employee theft or sabotage. This result is typical of 75 firms with similar plans surveyed by Senator Russell Long and the Senate Finance Committee in 1979.

Surveillance and Sanctions. Finally, of course, good management will make use of the same deterrent methods used by society at large in the effort to minimize crime. Good organization and record keeping will help to ensure that theft is promptly discovered. Retail merchants can hire professional "shoppers" to provide periodic checks on the competence and conduct of their clerks and a good team of shoppers will readily detect failures to ring up purchases and the like. If valuables are hard to get at, difficult to remove and spirit away, if losses are quickly detected and closely investigated when they occur, and if one's individual actions are periodically and unpredictably subjected to special scrutiny, the ordinary person will be disinclined to take the risk.

Summary

Most honesty tests are constructed on what appear to be rather simple minded principles, with items that are transparent and easy to fake. Most of the tests I've seen will be passed by anyone willing to observe the

following simple rules: "Don't admit anything, endorse strong punishment for any misconduct, and pretend you think that almost everyone is
honest." Most of these tests are proprietary and their validities are supported by publishers' claims rather than by data. The limited data that
have been made available indicate a very high false-positive rate.

There is reason for believing that improved management practices
offer an alternative to psychometric selection of employees. Moreover,
some of the same management practices that work to reduce employee
theft should also effect increased productivity. This alternative also has not
been systematically researched and is not even mentioned in the debate on
honesty testing.

The OTA report provides a clear and comprehensive review of the
cited facts about honesty testing and reaches the appropriate cautionary
conclusions. The APA task force report, in contrast, while urging the
honesty testing industry to be more honest and open in future, refuses to
recommend a moratorium on the use of honesty tests until data are available to assess their social impact and, indeed, opposes any governmental
regulation. An environmental impact assessment of this problem suggests
that widespread use of even the best selection devices, well constructed
and appropriately validated, may have an adverse impact on millions of
innocent people. If a headache remedy or other drug has not been shown
by independent studies to have beneficial effect, and can be shown on
existing evidence to put significant numbers of users at risk, we acknowledge the right and the responsibility of government to keep it off the
market. Although not obliged, like our medical brethren, to subscribe to
the Hippocratic oath, I believe we psychologists should nonetheless strive
at least to *do no harm*. On the existing evidence, honesty testing is likely to
do considerable harm and, in my view, should be closely monitored and
regulated.

Chapter 17

THE FOURTH DEGREE: POLYGRAPHICALLY INDUCED CONFESSIONS

To obtain a confession where guilt is indicated is the purpose and ultimate goal of the deception [lie detector] test.... The instrument and the test procedure have a very strong psychological effect upon a guilty subject in inducing him to confess.

—C. D. LEE, *The Instrumental Detection of Deception*, 1953

We get better results than a priest does.

—JOHN E. REID, *New York Times*, November 21, 1971

She was a journalism student working as a reporter for the *Minnesota Daily* and she was angry. She had told me on the phone that she wanted to do a story on the lie detector business, but it was plain that she had lost her journalistic objectivity. She wanted to blast them. This young woman had applied for a part-time job in a retail store that required a polygraph examination for preemployment screening. "I was kind of intrigued; I thought it would be interesting." She was so furious that I thought she must have drawn one of those prurient polygraphers who takes advantage of the examination situation to peek under the mental skirts of female applicants. I was mistaken; she had received the standard preemployment treatment. She was angry because after it was all over (she had "passed" the test) she felt violated. She realized that she had told the examiner things she hadn't told her priest, that she had blurted out all her little secrets, trying to explain why that implacable machine had "shown a reaction" to this or that question.

> You come in feeling like an honest, decent person and then he starts telling you that you're showing a reaction to some question about drugs or stealing and you try to think of a reason, things you've done, maybe years ago. You get to feeling as if the all-important thing is to get the machine to say that you're all right. You feel as if the thing can almost read your mind but not really accurately; it exaggerates, gets things mixed up. And you feel that if you tell the man everything you can think of, everything you've ever done that you might feel guilty about, why then finally it might come out with a clean record.

An ordinary employment interview does not have such impact. It may be a mistake to attribute all of the effect to the mystique of the polygraph, however. After all, the ordinary employment interviewer does not ask the same questions: "Have you committed an undetected crime?" "Did you ever steal merchandise from a previous employer?" It is possible that a skilled, tough interviewer could use the pressure of a level gaze, a skeptical expression, and an expectant silence to produce similar results from the same sorts of people who find themselves babbling in the polygraph room. What sorts of people are these? Polygraphers report that they elicit "damaging admissions" from 75% of job applicants. This statistic tells us at once that we are talking about ordinary, average people; it is the ones who do not make damaging admissions who are apparently exceptional.

We have already seen that polygraph examinations are widely used to screen applicants for jobs with federal security and police agencies and also applicants to state and local police departments. As early as 1961, in Orlando, Florida, nearly 900 applicants were pruned by aptitude tests and other criteria down to 45 serious candidates who then were subjected to a polygraph interrogation. According to C. A. Romig, 75% of this select group admitted to having committed "serious undetected crimes."[1] From 1963 through 1971, the Kalamazoo Police Department gave polygraph examinations to some 520 applicants who, collectively, admitted to more than 4,500 larcenies, including 89 burglaries. More than 160 admitted a variety of sexual peculiarities ranging from indecent exposure to bestiality,

35 confessed to previous arrests, and 45 admitted falsifying their application forms.[2] A Chicago police detective, applying for a responsible position with the Minneapolis Police Department, admitted numerous instances of illegal intimidation of suspects. One of his favorite methods had been to drive the handcuffed suspect at speed along the freeway, passenger door ajar, threatening to shove him out unless he confessed whatever it was the officer wanted to hear. One wonders whether this detective learned from his experience with the polygraph that equivalent results could be obtained less violently.

Preemployment admissions like these are obviously significant. An applicant who admits to burglary, arson, assault, passing bad checks, chemical dependency—to list some of the offenses acknowledged by police candidates—should reasonably be turned away. But not all the "damaging admissions" made by job applicants should be taken this seriously, not unless we are prepared to conclude that 75% of Americans are criminals and unemployable. Turn anyone inside out and "who shall 'scape whipping?" B. F. Skinner, a distinguished American psychologist, tells in his autobiography of running next door from the shoe store where he worked as a boy to buy ice cream with pennies taken from the till; this amounts to "theft from previous employer." Skinner also mentions incidents of embezzlement, sexual misconduct, and alcohol abuse, all in the course of the relatively sedate, privileged, and productive young manhood of an outstanding scientist.[3] I am not so distinguished as Professor Skinner, but I could a longer tale unfold of juvenile sins. So, indeed, could almost anyone; examine your own conscience. As suggested earlier in this book, one is inclined to wonder about those people who do *not* make "damaging admissions." Some undoubtedly are truly saintly; others have sense enough not to be bamboozled by the polygrapher and his machine and are self-disciplined enough to keep their own counsel. What about the real rogues, the deliberate villains who plan to rip off everything they can if they can get the job? Do they make "damaging admissions," or do they bluff their way through this situation as they have learned to do in others?

For many polygraphers, the posttest interrogation and the confession it so often induces is the real object of the whole exercise. A lie test diagnosis is a promissory note, not negotiable in many places, but a confession is pure gold, admissible in court, the finish to a criminal investigation, the commodity the client will be happiest to pay for. The Michigan State Police were among the earliest in this field, and in 1943, LeMoyne Snyder reported on the results of the use of the polygraph in some 900 criminal cases over a period of eight years.[4] More than one-third of these were solved by confessions obtained during or subsequent to lie detector tests. The Los Angeles Police Department maintains a polygraph laboratory with several full-time examiners, who estimate that they obtain confessions from about 25% of the suspects examined. More specifically, about 60% are diagnosed deceptive (30% truthful and 10% inconclusive or interrupted) and some 40% of that 60% produce confessions.

Criminal investigation involves two equally important—and often equally difficult—aspects: (1) identifying the guilty person and (2) obtaining admissible evidence against that person sufficient to obtain a conviction. A valid confession solves both of these problems, neatly and inexpensively. Any technique that can produce a confession from a fourth to a third of those cases in which suspects are apprehended and agree to be tested is an undoubtedly valuable tool.

Certain problems remain to be considered. First, who is it that confesses? There are no statistics to guide us, but it seems most unlikely that the sophisticated professional criminal will be stampeded into a confession by anything a polygrapher might tell him. Many felons, however, are not professional criminals, just victims of passion or impulse or circumstances who never expected to get into such a mess and do not expect to get away with it.

Second, is the polygraph really necessary? A recent news story described how one ingenious pair of detectives used a photocopier to obtain a confession from an unsophisticated suspect. Each time the man failed to produce the answer they wanted, the machine would groan and whir and deliver the printed message, "He's lying!" copied from a previously inserted master. Most urban policemen know the trick of wrapping the squad car's microphone cable around the arm of a gullible suspect and then surreptitiously touching the transmit button when he denies their accusation. "You see that red light, George? That red light means you're lying to me. Let's try it again now." Are any of these stage props actually necessary? Are they more important for breaking down the culprit's defenses or for enhancing the interrogator's confidence and aggressiveness? If the St. Paul police use the polygraph, while those in Minneapolis rely on old-fashioned straight interrogation by experienced detectives, will St. Paul produce a higher proportion of confessions? The experiment has not been tried, and it is difficult to guess the outcome. Perhaps most of that feckless 25% who confess under polygraphy would confess anyway with the help of a little intelligent questioning.

Third, every lie detector test involves deception, or attempted deception, of the suspect, and confessions induced by this procedure are to that
extent tainted. Tainted or not, one could argue that a valid confession
obtained through trickery is better than no confession at all. There is
something offensive about the fact that the ignorant and the gullible are
most vulnerable to such devices; educated people with good legal counsel
seldom make confessions under these conditions. But the key again is
whether the confession is valid. Foolish and gullible people who break the
law are more liable to get caught, but society does not owe a stupid man
protection in order that his chances for success in crime equal those of
someone better endowed. The real problem about depending on the myth
of the lie detector for inducing confessions is that this myth must gradually
be eroded. A young policeman may reasonably use CVSAs (voice stress
analyzers; see Chapter 11) and photocopiers as long as they work, but he
should, meanwhile, learn to interrogate, against the day when the tricks
no longer work and he is on his own.

Finally—and most importantly—can we assume that polygraphically
induced confessions are always valid? The answer is that we cannot, not as
long as some police give too much rein to their amateur notions of psychology and play upon the susceptibilities of the gullible. Just after the first
edition of this book came out, a colleague sent me a clipping from the *New
York Times* headed "Rejected Confession Raises Questions on Lie-Detector
Use."[5] Matthew Johnson, an 18-year-old high school student, had confessed to the stabbing of Renee Walker in her apartment where Johnson's
uncle was the building superintendent.

> Mr. Johnson, who had never been arrested before, said he was "ex-
> hausted" by 11 hours of intermittent questioning by detectives and by
> three lie detector tests. Testifying at a hearing … he contended that
> after the tests, a detective warned him, "You are really lying. We can't
> stay here all day, playing games with you." "The cops kept asking me
> questions," Johnson said, explaining why he confessed to a detective
> and later made admissions on videotape to an assistant district attor-
> ney. "I couldn't take it no more."

Johnson was held in custody on Rikers Island for 17 months before a state
supreme court justice in Manhattan ruled that his confession had been
coerced and could not be used as evidence against him. There being no
other evidence implicating Mr. Johnson, the justice ordered the youth
released from prison.

Most lawyers know of Borchard's *Convicting the Innocent,* in which are reviewed 65 cases of persons imprisoned or executed on the basis of confessions proved later to be false.[6] According to Richard Leo,[7] "False confessions have long been one of the leading causes of miscarriages of justice in America." A shocking and substantial literature has appeared in support of this claim since the first edition of this book was published.[8] Most of the long list of cases aggregated by Leo and the other cited authors are of false confessions elicited by police interrogation without the help of either physical abuse or the polygraph. The main psychological principle on which modern interrogation methods are based is simple: Persuade the suspect that you know he is guilty, that no one will believe his denials given the facts allegedly in evidence, and that his best option is to confess and cut the best deal that he can. The addition of the polygraph rationalizes the essential postulate: "Now that this impartial, scientific instrument has detected your deception, no one will believe your denials." But the polygraph sometimes goes even further by persuading the innocent suspect to believe the lie detector against the evidence of his own memory.

The Peter Reilly Case

A vivid illustration of how a false confession is born can be found in Barthel's *A Death in Canaan*, a true account of a Connecticut murder investigation in 1974.[9] Peter Reilly, 18 years old, came home one night to find his mother's mutilated body on the floor of the bedroom, her last gasps of breath bubbling from a severed windpipe. The police whom he summoned placed Peter in a squad car, where he shivered alone for three hours while detectives rummaged inside the house. After a few hours' sleep at the state police barracks, Peter was given a lie detector test. The pretest interview, the test itself, and the subsequent hours of interrogation were tape-recorded, so we can follow what actually happened. During the lie test, Peter gave strong physiological reactions to questions like "Last night did you hurt your mother?"—hardly surprising under the circumstances, but these policemen had taken classes in polygraphy that apparently corroded their common sense. "Pete, we go strictly by the charts. And the charts say that you hurt your mother last night." The transcripts reveal how Peter, shocked and distracted, wanting to cooperate with the kindly policemen, let himself come to believe in "the charts" rather than in his own memory:

"The test is giving me doubts right now. Disregarding the test, I still don't think I hurt my mother."

Sergeant Kelly explains that Peter's mother really had it coming; she was always on his back, until he finally blacked out and killed her and now:

"You're so damned ashamed of last night that you're trying to just block it out of your mind."

And Peter:

"I'm not purposely denying it. If I did it, I wish I knew I'd done it. I'd be more happy to admit it if I knew it. If I could remember it. But I don't remember it."

Peter had some doubts at first about the machine.

"Have you ever been proven totally wrong? A person, just from nervousness, responds that way?"

But Sergeant Kelly was a rock of certainty.

"No, the polygraph can never be wrong, because it's only a recording instrument, recording from you."

"But if I did it, and I didn't realize it, there's got to be some clue in the house."

"I've got this clue here [the charts]. This is a recording of your mind."

"Would it definitely be me? Could it have been someone else?"

"No way. Not from these reactions."

There was a problem about the legs, both heavy femurs broken just above the knees. Peter hadn't known his mother's legs were broken.

"Did you step on her legs or something? While she was on the floor? And jump up and down?"

"I could have."

"Or did you hit her?"

"That sounds possible."

"Can you remember stomping on her legs?"

"You say it, then I imagine I'm doing it."

"You're not imagining anything. I think the truth is starting to come out. You want it out."

And then finally, after several hours,

"Well, it really looks like I did it."

Thus, while the real murderer was effecting his escape, Peter Reilly was signing a confession to the mutilation slaying of his mother, and it was two

years before that confession was shown to be false, that Peter could not
have committed the crime, two years during which the trail left by the
actual murderer, still at large, grew faint and cold.

Reading the transcript of the Peter Reilly interrogation, any sensible
person should be able to recognize at once that this confession was meaningless, an eruption from seeds planted one by one in the mind of an
exhausted and impressionable boy. The police, eager to "solve" their case
and blinded by a perverse and groundless faith in the polygraph, apparently were satisfied they had their killer. How the district attorney could
have agreed to prosecute such a case is difficult to understand. The Peter
Reilly story should be required reading for all polygraphers, every prosecutor, every juror in cases where repudiated confessions figure in the
evidence.

Why Do People Confess?

The etymology of "third degree," an American colloquialism going
back to the 19th century, is somewhat obscure. It refers to protracted and
exhausting interrogation accompanied by threats, intimidation, or actual
torture. The "sweat box" was a popular technique of administration, a
small cell set next to a furnace.

> A scorching fire would be encouraged in a monster stove adjoining,
> into which vegetable matter, old bones, pieces of rubber shoes, and
> kindred trophies would be thrown; all to make a terrible heat, offen-
> sive as it was hot, to at last become so torturous and terrible as to cause
> the sickened and perspiring object of punishment to reveal the inner-
> most secrets he possessed.[10]

According to this same author, "third degree" is an allusion to Masonic
rites and to the trials and rituals administered to those aspiring to the Third
Degree, or Master Mason level, of that hierarchy. "The officer of the law
administers the 'first degree,' so called, when he makes the arrest. When
taken to the place of confinement, that is the 'second degree.'"

We need not inquire why the third degree is conducive to confession,
valid or otherwise, but it may be useful to consider why polygraphic
interrogation—the fourth degree—without benefit of sweat boxes or
bludgeons is nonetheless so frequently effective. "All men are liars," says
the psalmist, but most of us play the game according to established rules,
and one basic rule is that when the jig is up, you should resign. When your
inquisitor knows the truth, when your story has been shown to be incon-

sistent or incoherent, when the counterevidence is obvious before you, then you are supposed to admit defeat. Other rules are that one lies only for an important reason, and preferably never to a friend. A good interrogator becomes your friend, helps you to justify whatever it was that you did, making it easier for you to acknowledge the truth. Lying is difficult, tiring. A good examiner makes you see how much easier it is, what a relief it will be, to stop the hopeless battle of wits and get it off your chest. He speaks not of punishment, but of reconciliation. Most important of all, he tries to make you realize that your story won't hold water, that it doesn't square with the facts or is inherently incredible. He may allude to other witnesses, other evidence, information in the light of which your denials are transparent and futile. He is careful from the outset to make you feel that your objective is to make *him* believe you and that if you cannot win him over, then the game is lost. He wants to prevent you from stubbornly repeating the same story, "take it or leave it," because once you have ceased to care whether he believes you, then his leverage is lost. Some people, including many experienced criminals, never expect the examiner to believe them, and this pessimistic attitude protects them from much of the stress of interrogation. This is a rational attitude, since normally it does not matter what the examiner believes, only what he can prove. But most of us, when we are lying, act less realistically, as if our denials were mere paper shields that the examiner's disbelief can shred like a knife, leaving us exposed.

Once he has you playing his game, has you trying to convince him, a good interrogator will devise a variety of methods to persuade you that you have finally lost, that it is hopeless, he has seen through you, and you might as well confess. The polygraph is only one such method, but it is an easy one to use and frequently effective. "We go strictly by the charts, and the charts say you hurt your mother last night." If Peter Reilly had actually been guilty, how could he have stood up to that? Yet, why should anyone care what "the charts say"? Has man evolved some innate tendency to defer to oracles? Or is there some unconscious logic that informs us that if the lie detector knows, then everyone will know, and that a lie that no one believes is useless?

Confessions and the Courts

Because police are often overeager in their attempts to elicit confessions, American courts have reacted by application of the Exclusionary

Rule, suppressing or refusing to admit into evidence confessions obtained by illegal means or under circumstances that the court deems to be coercive so that the resulting confession cannot be considered to have been "freely and voluntarily given." Judges have begun to speak of "psychological coercion," a concept so broad as to include potentially all the tactics of the skilled interrogator that we have considered above. The only confession that would be wholly untainted by psychological coercion might be one obtained from a citizen who walks in off the street and volunteers his story. We abjure torture, threats, and intimidation because (1) such treatment of criminal suspects, some of whom are innocent, is uncivilized and violates principles of human rights that our society holds dear, and because (2) such forms of coercion are bound to elicit a high proportion of false confessions. But once we start to talk about psychological coercion (and we have already started when we deplore threats and intimidation), where do we draw the line? Any action, any external stimulus that increases the probability of a confession's being emitted by a suspect, would constitute some degree of "coercion" under this broad definition, and the voluntariness of the confession would be to some degree diminished. Merely asking, "Did you do it?" is coercive in this sense because presumably some persons require at least this much external stimulation or help before acknowledging their guilt. To establish a friendly, confiding relationship with the suspect, to discuss the advantages of getting his guilty secret off his chest, to suggest that his alibi appears incredible are tactics that are more coercive still and will yield confessions that are in some sense less than completely voluntary. To use trickery—to allude to nonexistent witnesses or other imaginary evidence, to pretend that confederates have already confessed, or to indicate that the polygraph charts have somehow revealed the truth—will further increase the proportion of confessions obtained. Such tactics do not seem to me to violate the first canon implied above: Such methods are not intrinsically harmful or inhumane. However, trickery and the confessional pressures of the polygraph examination do greatly increase the dangers of eliciting false confessions.

I was once consulted by a young woman who had signed a false confession to the theft of money from the store where she worked. The corporation's chief security officer, a man trained in interrogation techniques by the Army, had scorned the polygraph for this $10 matter and had relied instead on standard methods. He indicated that what she allegedly had done was not really so bad: "Everybody slips once in a while." But, at the same time, he exaggerated the magnitude of the charge so that she might more readily confess to a lesser one: "How much do you think you've taken altogether? $1,000? $500?"—although the sum he really had in mind was less than $50. He made her see the futility of continued denial: "I saw you take some money out of the register with my own eyes, Carol," although the view from his distant hiding place was so indistinct that he could not have been sure what he saw. And he contrasted the minor consequences of confession with the dire consequences of obstinacy: "If you sign this statement, we can settle this now between us. Otherwise, of course, I'll have to turn the matter over to someone else." Assuming he meant the police, Carol ultimately signed even though she was innocent.[11]

The Embassy Marine Guard Scandal

In 1986, newspaper headlines revealed that Marine guards at the U.S. Embassy in Moscow had been found to have conducted guided tours for Soviet KGB agents through the secret inner sanctums of that building. A *Time* magazine cover graphically portrayed by far the worst shame ever to bedraggle the honor of the Corps. Several silent months after these horrific revelations, an article by *Washington Post* reporter Don Oberdorfer revealed the even more shameful truth. Agents and polygraphers of the Naval Investigative Service (NIS) learned that a Native American Marine, Sgt. Clayton Lonetree, had befriended a Russian woman employed by the KGB. They then had sought out three other enlisted Marines who had also worked as guards at the U.S. Moscow Embassy during Lonetree's tenure there and subjected them to repeated cycles of polygraph testing and interrogation. Each polygraph test included increasingly bizarre allegations to which these young Marines reacted with increasing physiological disturbance. Ultimately, Cpls. Arnold Bracy and Robert Williams and Sgt. Vincent Downes signed wildly incriminating statements—statements they at once repudiated after being rescued from the NIS interrogators. According to Robert Lamb, head of the State Department's Diplomatic Security Bureau, "there were things in Bracy's statement that could not have happened."[12] These were young African American Marine noncoms, plucked from their subsequent posts by NIS investigators and questioned, more or less nonstop for three days, each successive polygraph test suggesting still more outlandish possibilities, accusations that the young men reacted to with increasing alarm, thus confirming the polygraphers' beliefs that they were on the track of something big. Sometime later I received a phone call from a Marine colonel, a Judge Advocate General officer who had served as defense counsel in Cpl. Bracy's court-martial. This colonel wanted nothing more from me than understanding corroboration of his outrage at what these NIS operatives and their "damnable polygraphs" had done to his client and, especially, to the reputation of his beloved Marine Corps. Reagan administration officials finally admitted that, in fact, the Marines didn't admit any Soviet agents into the embassy. As journalist Oberdorfer wrote, "the government has been grappling mainly with phantoms of its own invention."

Another type of false confession results when the suspect is actually led to believe that he is or might be guilty. In the Peter Reilly case we can trace this "brainwashing" process from first to last. However, I have seen transcripts of confessions obtained in the course of a polygraph test that reveal with equal clarity that the confession is valid. The suspect volunteers facts and details of his crime not suggested to him by the examiner. The confession is not a mere acknowledgment of guilt but amounts rather to a narration of a coherent and at least partially verifiable description of the particulars of the event. As someone not learned in the law, I would think it reasonable to give credence to confessions of this latter kind even though obtained partly through trickery, including the deception inherent in a polygraph examination. But the problems of interpreting the Fifth Amendment and of establishing rules to ensure adherence to those interpretations are deep matters and beyond my competence.

To sum up, polygraphic interrogation in the hands of a skillful examiner is a powerful cathartic (emetic?), an effective inducer of confessions. Its confessionary influence may be most effective with the naive and gullible or, among criminals, with the less experienced, less hardened types. In preemployment or other screening applications, it appears that the majority of ordinary citizens may be led to make damaging admissions in this secular confessional. In criminal investigations, as many as 25% of those cases where suspects are available for questioning may be settled expeditiously by confessions obtained through polygraphy. This aspect of polygraphic interrogation is quite independent of the actual validity of the technique as a detector of deception. If all polygraphs were stage props it is likely that just as many admissions or confessions would be elicited. Certainly much of the popularity and utility of the polygraph derives from this incidental effect. This may be why so many polygraphers show little interest in research on the actual validity of the various forms of polygraph test; even if its true validity is no better than chance, so long as most people believe in the lie detector or the voice stress analyzer, these tools will continue to elicit admissions and confessions, and that is their principal purpose. Because they are so effective, however, these methods commonly inflict great stress and emotional disturbance on the innocent and guilty alike. The fourth degree may leave no cuts or bruises, but it hurts—that is why it works. And because it works so well, one should distrust any confession obtained by modern interrogation methods, whether the polygraph was employed or not, unless that confession can somehow be confirmed.

Chapter **18**

THE LIE DETECTOR
AND THE COURTS

As a tool of persuasion and advocacy, the polygraph stands alone because of its ability to slash the jugular vein of each case which must ultimately turn on whether one or more persons are attempting deception. If the credibility of the defendant can be established or destroyed in a criminal case, or if the plausibility of either party in a civil case can be demolished, there is nothing more to settle, nothing left for judge or jury except the assessment of penalty or awarding of damages. No other courtroom tool possesses this awesome power.

—MARSHALL HOUTS, LL.B.[1]

When polygraph evidence is offered ..., it is likely to be shrouded with an aura of near infallibility, akin to the ancient oracle of Delphi.

—United States v. Alexander, 8th Circuit Court of Appeals, 1975

"If the law supposes that," said Mr. Bumble..., "The law is an ass, an idiot."

—CHARLES DICKENS, *The Pickwick Papers*

The Frye Case

In November of 1920, Dr. R. W. Brown was shot to death in Washington, D.C. The following summer a young black man, James Alphonzo Frye, was arrested and grilled for several days by the D.C. police. Frye finally admitted the Brown murder but repudiated this confession just before his trial, claiming that he had been promised half of the $1,000 reward if he would falsely confess to the killing. Our old friend, Dr. William Moulton Marston, administered his blood pressure lie test to the defendant and concluded that he was innocent. The defense petitioned the court to allow Dr. Marston to be qualified as an expert witness and to present to the jury the results of his lie test. Judge McCoy, presiding, excluded this evidence and his ruling was later upheld by a federal appeals court in language that for nearly half a century stood as a barrier between the lie detector and the courtroom:

> Just when a scientific principle or discovery crosses the line between the experimental and demonstrable stages is difficult to define. Somewhere in this twilight zone the evidential force of the principle must be recognized, and while courts will go a long way in admitting expert testimony deduced from a well-recognized scientific principle or discovery, the thing from which the deduction is made must be sufficiently established to have gained general acceptance in the particular field in which it belongs. We think the systolic blood pressure deception test has not yet gained such standing and scientific recognition among physiological and psychological authorities as would justify the courts in admitting expert testimony deduced from the discovery, development and experiments thus far made.[2]

Ironically, Frye, who was found guilty of second-degree murder and sentenced to life imprisonment, was subsequently exonerated and set free. Marston had been right after all.

In recent years the barrier presented by the *Frye v. United States* decision has been eroding. At least 17 states now admit the results of stipulated lie tests, polygraph examinations administered after prior agreement by both sides.[3] The rationale here, apparently, is that anything is fair that both sides agree to, although, as Attorney Lee M. Burkey has pointed out:

> It is difficult to understand how the polygraph method is improved merely because the parties stipulate to be bound by it. Would the court approve a stipulation to be bound by the toss of a coin?[4]

Law professor Faigman makes a similar point:

> To be sure, parties regularly stipulate to evidence. But polygraphy is unique, in that the stipulation occurs before the real evidence—the polygraph result—exists. If polygraphy is not reliable, this stipulation amounts to a gamble that courts might be reluctant to endorse.[5]

Allegations of Sexual Abuse

Allegations of sexual abuse—adult rape or especially the sexual abuse of children—present courts and the police with an especially difficult problem. Typically the physical evidence is inconclusive or nonexistent and sometimes the alleged victim may be motivated to dissemble or, in the case of very young children, the accusations may be inadvertently planted in their minds by incautious and suggestive interrogation. These are the kinds of cases in which one wishes that there were a Truth Verifier.

One of my first experiences as an expert witness for the defense saw truth triumphant without any help from me. Ed and Marilyn W. had gone through an acrimonious divorce, the pros and cons of which I am not privy to. Ed subsequently remarried a woman named Jane who was a widow with two youngsters by her first marriage. Ed and Marilyn had a 10-year-old daughter, Julie, who spent a few days each month with her father and his new family. A few months before my appearance in Cedar Rapids, Marilyn and Julie had visited the county attorney with hair-raising charges. During Julie's last visit to her father's home, she had been forced to witness him and his new wife in sexual congress and, later, had been sexually abused in bed by both of them! Ed had a good reputation in the community but Marilyn was widely considered to be unreliable. However, these were serious charges and the county attorney felt obligated to take some sort of action. The Cedar Rapids police had a polygraph examiner and it seemed natural to turn to him for help. "Ed, if you'll agree to take a lie detector test on this, I will promise to drop the charges if the polygraph says you're telling the truth. But you have to agree in advance that we can use the evidence against you if you fail the lie test." Ed promptly agreed, took the test, and failed it. That was when I was invited to drive down to Iowa.

During the prosecution's case-in-chief, the most important witness was little Julie W., the alleged victim. She was a good witness, cool and self-contained. She told her damning story clearly and without hesitation or prompting. I think everyone in that courtroom, except perhaps for Ed and Jane, were a bit teary-eyed when Julie had finished her direct examination. No one envied the defense attorney's task when he got up to cross-examine. His expression was guileless and concerned. "Julie, did I understand that this was your tenth birthday when all these awful things happened?" "Yes, it was." "And didn't they do anything nice for your birthday, no presents or anything?" "No, just those dirty things." "Julie, I'm going to show you a picture that was taken that evening." (He hands her an enlargement of a snapshot with a copy to the judge and another to the prosecutor. The home photograph shows Julie, smiling, wearing a paper party hat, about to blow out the ten candles on a birthday cake. In the background is Ed's new wife, all smiles, with her two children, and the table is piled with wrapped birthday presents.) Julie looked at the photograph for what seemed like a long time, then laid it down, looked up defiantly, and said very clearly, "Okay, I lied." I never had to testify at all.

Francine Bronson (not her real name), a nurse in Yakima, Washington, was charged by the new wife of her ex-husband of sexually abusing her own four-year-old son. Because there was no real evidence of abuse, the district attorney offered Nurse Bronson a deal like Ed's: "If you can pass a polygraph test, we'll forget this thing but you have to agree in advance that, should you fail the test, we can use that result in evidence against you." Frantic to be freed of this outrageous allegation, Nurse Bronson agreed—and failed the test. The polygraph examiner testified that he had graduated from a recognized polygraph school (not mentioning that the entire course of instruction lasted only eight weeks) and that he had given hundreds of tests in the course of his career. He said that his determinations had never yet been proven to be wrong. He explained that he had administered a state-of-the-art control question polygraph test to the defendant and that, in his professional opinion, she was clearly deceptive in denying having sexually abused her child. Called to testify for the defense, I thumbtacked the defendant's polygraph chart to an easel in front of the jury box together with the numbered list of the questions she had been asked. I showed the jurors how the examiner had marked the chart in pen where he had asked the numbered questions and that most of them were followed by changes in Nurse Bronson's blood pressure, in her breathing patterns, and in the sweating of her palms.

Here is where the examiner asked her "Have you ever committed an unusual sex act?" and you can see that there was some reaction to that question. But here, where he asked, "On the date of May 14, did you take Johnny's penis in your mouth?" that was followed by a much larger reaction. As I explained earlier, you "fail" a polygraph test if you are more disturbed by the relevant question than you are by the control question. As you can see, Nurse Bronson was clearly more bothered by the accusation that she took her little boy's penis in her mouth than she was by the question about "unusual sex acts." That is why—that is the only reason why—the examiner concluded she was lying about abusing little Johnny.

I was standing close enough to see the jurors' eyes widen with shocked understanding at this news. And I was not surprised to learn that it had taken them all of an hour and a half to bring in a verdict of not guilty. But not every defendant in Nurse Bronson's situation has an attorney dedicated enough to bring in a competent rebuttal witness. And the expense, not to mention the emotional cost, of this trial was assuredly money for which Yakima County (and Nurse Bronson) had better uses.

The Polygraph for the Defense

Another development flew more directly in the face of *Frye* and was largely a result of the efforts of the celebrated defense counsel F. Lee Bailey. During the 1980s, Bailey and the former Hollywood polygrapher Ed Gelb figured in a weekly television program, *Lie Detector*, in which participants wishing to demonstrate their innocence of certain allegations were tested by Gelb and the results announced to the waiting world. Himself a former professional polygrapher, Bailey advocates the principle that the defendant should be allowed to "prove his innocence" with evidence of having passed a lie test, introduced even over the objections of the prosecution. It was no doubt at Bailey's urging that O. J. Simpson was given a private polygraph test, which he presumably failed since Mr. Bailey never offered it in evidence at trial. California and Massachusetts both permitted introduction of the results of exculpatory or "friendly" polygraph tests for a few years, but this practice was ended in California in 1983 by action of its legislature and the Supreme-Judicial Court of Massachusetts put a stop to it in that state in 1989. New Mexico is presently the only state that continues, by statute, to endorse this principle.[6]

However, the U.S. Supreme Court's decision in *Daubert v. Merrell Dow Pharmaceuticals*,[7] while it did not consider polygraph evidence specifically, changed the rules concerning the admissibility of disputed scientific evidence. Acknowledging that the *Frye* test had been superseded by the legislatively enacted Federal Rules of Evidence, the Court explained:

> That the Frye test was displaced by the Rules of Evidence does not mean, however, that the Rules themselves place no limits on the admissibility of purportedly scientific evidence. Nor is the trial judge disabled from screening such evidence. To the contrary, under the Rules the trial judge must ensure that any and all scientific testimony or evidence admitted is not only relevant, but reliable.[8]

A consequence of *Daubert* has been that many federal courts, and some state courts following the federal lead, have been unwilling to per se exclude proffers of polygraph test results over the objection of one party. Instead, trial judges have been holding evidentiary hearings to determine whether the proffered evidence meets the new *Daubert* criteria, the first and most important of which is

1. "Whether a theory or technique ... can be (and has been) tested [since] ... this methodology is what distinguishes science from other fields of human inquiry."[9]

Three other *Daubert* criteria, each considered to be relevant though not dispositive, are these:

2. "Whether the theory or technique has been subjected to peer review and publication."
3. "The court should consider the known or potential rate of error and the existence and maintenance of standards controlling the technique's operation."
4. "Finally, 'general acceptance' [within the relevant scientific community] can yet have a bearing on the inquiry.... A known technique that has been able to attract only minimal support with the community ... may properly be viewed with skepticism."[10]

Earlier (in Chapter 4), we saw that a logical consequence of a highly accurate Truth Verifier would be the elimination of the jury in most criminal trials, replaced by a polygraph examination of the defendant. While most modern polygraphers are eager to be qualified as expert witnesses in court, none of them publicly advocates so radical a step as the substitution of a polygraph machine for the jury box as a standard courtroom fitting. Yet this modesty seems inconsistent inasmuch as polygraphers claim to be able to arrive at the correct diagnosis 95% or 99% of the time; no one seriously believes that judges and juries are as prescient or as accurate as that. If David Raskin or Ed Gelb have examined the defendant, and if the findings involve at most a few percent error as they claim, then in the interests of justice—not to mention efficiency of adjudication—the court should hear Mr. Gelb or Dr. Raskin first and then issue a directed verdict in harmony with their conclusions.

But we have also seen that, in fact, the lie detector test is *not* 99% or 95% or even 90% accurate and that there is strong reason to doubt that any lie test will ever attain such levels of accuracy, based on what we have long known about the complexity and variability of the human animal. Both the Truth Control Test (which requires that the subject believe he is suspected of an equally serious but actually fictitious crime) and the genuine Control Question Test (which requires that we know the subject to be guilty of an equally serious crime—but that he not know that we can prove his guilt) might in principle be highly accurate. But both of them require a complicated deception of the subject and are invalidated if the deception does not work. If courts began deciding cases on the basis of these tests one can be sure that defense attorneys would quickly learn the theory of the procedures and that the necessary deceptions would become impossible.

We have seen, in Chapter 12, that most members of the relevant scientific communities agree with the evaluation of these lie detection methods presented in this book and most, because of the lie detector's dubious validity, its vulnerability to countermeasures, and its subjectivity and bias, oppose its use as evidence in court. The forms of polygraph examination actually used in criminal investigations, the clinical lie test of the Reid and Arther schools or the various forms of polygraphic lie test such as the CQT or the DLT, have an average accuracy that is certainly less than 75%, according to the best data available, and these tests are strongly biased against the truthful respondent. Basing criminal verdicts entirely or substantially on such evidence would be wantonly unfair and improper. On the other hand, every day criminal court juries listen to evidence that is substantially less than 90% accurate on the average; eyewitness testimony is notoriously undependable. Other types of expert witnesses, similarly, are not held to such demanding standards of near infallibility. My own profession used to provide perhaps the worst example. On questions of mental status, competency to stand trial, and the like, one got the impression that both sides could always find psychologists or psychiatrists of equal status and credentials who would give exactly opposite "expert" opinions. Publication of the more recent revisions of the official Diagnostic and Statistical Manual, DSM-III and now DSM-IV, has moderated these battles of experts by providing explicit, reasonably objective criteria that make psychiatric diagnosis much more reliable, if not more valid. Handwriting and "voiceprint" identification are imperfect art forms and yet juries are permitted to hear the opinions of such experts and make of them what they will. Even ballistics and fingerprint identification are fallible, as is proven every time the defense produces someone who reads the signs differently than did the prosecution's expert. Why should the polygrapher be excluded from this parade of imperfectibility? Research indicates that the Rorschach inkblot test is probably less valid than the Control Question Test; why then should a psychologist who has administered a Rorschach be permitted to state his opinion about the defendant's competency, while a polygrapher, who has administered a CQT, is not allowed to testify as to the defendant's veracity?

Two separate lines of argument can be adduced in justification of this seemingly arbitrary discrimination. One has to do with the apparent paradox that the same lie test that averages (say) 70% correct conclusions on the general run of cases may be only 50% accurate (or worse) on the selected subset of cases that will be offered into evidence in court. The other line of argument contends that the polygrapher *is* different from other experts in that his testimony is "improperly preemptive of the act of judgment of the (trier of fact) on the issue of credibility."[11] Let us consider these two arguments separately and in more detail.

The Question of Base Rates

In psychometric parlance, the *base rate* of a condition (such as schizophrenia or being guilty of a certain crime) refers to the frequency of that condition among the members of a specified population. In the Horvath study of the accuracy of the CQT, the base rate of liars in the population studied was intentionally set at 50% by selecting for analysis polygraph charts from criminal suspects half of whom were later determined to be guilty and half innocent. In the Barland and Raskin study, an unselected series of cases was employed of which 78% turned out later to be guilty according to the criterion used; the base rate for lying here was 78%.[12]

What is the base rate of lying among the defendants whose stipulated polygraph tests are admitted into evidence in American courts? In my experience, prosecutors offer a defendant the option of stipulating to a lie test only in situations where the prosecution's case is weak and unlikely to sustain a conviction. Under these circumstances, if the defendant passes the lie test, charges can be dropped and nothing lost. But if the defendant fails the lie test, then the prosecution can proceed with a much stronger case and better chance of winning. Therefore, in the vast majority of those stipulated lie tests that are actually offered into evidence, the defendant *has* failed the test. Thus, we are dealing here not with criminal defendants in general but rather with that subset of defendants against whom persuasive evidence of guilt is not available. Among this select subset, the most likely reason that the prosecution's case is weak may be that the defendant is in fact innocent. For purposes of illustration, I shall assume that 70% of these particular defendants are innocent, which means that the base rate for lying among this group will be 30%. Keeping in mind that existing accuracy studies *must* overestimate CQT validity, the studies summarized in Table 8.2 suggest that 85% of liars will fail these stipulated tests (assuming they have not learned how to beat the polygraph) and that about 40% of the truthful suspects will also fail. That is, out of each 100 stipulated tests, 26 of the 30 guilty suspects will be classified deceptive and so will 28 of the 70 innocent suspects. Since only the failed lie tests will be presented to the court, 28 of the 54 lie tests offered into evidence on this principle, 52% of them, will be erroneous, a rate of accuracy actually worse than could be obtained by flipping coins!

When we think of criminal suspects against whom the state has difficulty obtaining adequate evidence, we are inclined to think of organized crime figures, professionals with clever lawyers. But those same clever lawyers will not permit their clients to accept the prosecution's offer of a stipulated lie test. The defendants we are now discussing are likely to be small-time or first-time offenders, if they are guilty, and more likely still to be innocent victims of circumstance. If my estimate of 30% guilty is too low, then a larger fraction of the lie tests admitted into evidence will be valid, perhaps half of them or more. Conversely, if the state's polygrapher is more like the ten examiners studied by Horvath than like the two Ph.D. examiners involved in the Barland and Raskin study, the stipulated lie tests presented to the court will have still lower accuracy. We now have an answer to lawyer Burkey's question about "how the polygraph method is improved merely because the parties stipulate to be bound by it": It is not improved but it *is* changed. The fact of stipulation means that courts will see a selected group of lie tests the average validity of which will be substantially poorer than is true for lie tests in general. On the reasonable assumptions employed in the above example, this group of lie tests will have a probative value of about zero.

Suppose instead that the courts accede to F. Lee Bailey's proposal and permit the defense to offer lie test evidence over the objection of the prosecution. The humane intention here would be to compensate for the greater resources available to the state by giving the defendant this opportunity to prove his innocence. We must now imagine every defendant shopping for a friendly polygrapher in the hope of achieving a "pass" that could be offered into evidence—and that no defendants would offer testimony of a failed lie test. That is, we are now talking about criminal defendants against whom the evidence is strong enough for the prosecution to bring them to trial, rather than the subgroup against whom the prosecution's evidence is weak. If we assume that 80 out of each 100 criminal defendants actually brought to trial are in fact guilty (most officers of the court would set the figure higher), then about 12 of the guilties and 12 of the innocents should present evidence of passed lie tests, evidence that will be wrong 50% of the time. This unhappy result is based on the assumption that lie tests given privately to Mr. Bailey's clients are as accurate in detecting lying as those given under adversarial conditions. We shall see in the next section that these "friendly" polygraph tests are in fact more likely to be passed by guilty defendants. This chance level of errors also assumes that none of the guilty defendants who can afford defense attorneys in Mr. Bailey's class will have learned—in Chapter 19 of this book or in diverse other ways—how to beat the lie detector even though guilty. Because both assumptions are so dubious, it is likely that admission of exculpatory polygraph test results into evidence at trial will have a probative value that is *negative*.

As a final illustration, consider the use of lie detectors in preemployment screening. Polygraphers report that some 75% of job applicants make damaging admissions in the course of such screening, admissions offered to clear their consciences in order to obtain a truthful verdict from the polygraph. We can assume that most of those who make damaging admissions do not then feel required to lie to the polygraph, and that at least some applicants have nothing either to admit or to lie about. If only 20% of job applicants try to lie during the actual polygraph test, then about 490 out of each 1,000 applicants should fail the test and 65% of them will be false-positives, truthful victims of lie test errors.[13]

The base-rate problem we have been discussing has to be considered whenever the proportion of liars in the group actually being tested is appreciably different from 50%. In all three examples, each based on reasonable assumptions, asymmetrical base rates would be expected to lower the average validity of those lie tests in which we would actually be interested, for example, those lie tests that would actually be used in court. Given a polygraphic lie test with an average validity of 70%—the most optimistic estimate we can make on the research available—the base-rate problem reduced the effective validity in all three examples down to around chance levels or even worse. If the courts attempted to minimize the base-rate effect by requiring all defendants to take polygraph tests and then admitting the results into evidence whether pass or fail, they would still face the irreconcilable difficulty that about half of all innocent defendants tend to be erroneously diagnosed as deceptive.

The "Friendly Polygrapher"

Dr. Martin Orne has pointed out that the well-established tendency for truthful subjects to fail the lie test may not hold in the special situation where the polygrapher has been engaged by the respondent or his attorney and is in that sense "friendly" to the respondent's interests.[14] The examiner will have a natural inclination to serve his clients' interests. When the respondent on his own initiative requests and pays for a polygraph test, the examiner will be disposed to expect a truthful result. An attorney who hopes for a truthful result will be more likely to make subsequent referrals to an examiner who produces this desired outcome. In most cases, only by producing a truthful diagnosis can the examiner hope to earn subsequent witness fees.[15] For all these reasons, even the most ethical polygrapher will approach such a friendly examination with both the desire and the expectation that the respondent will "pass" the test.

Many objective psychological tests are relatively immune to the hopeful anticipation of the test administrator. But the lie test contains large elements of subjectivity. The polygrapher constructs the test questions to suit the particular case; his manner will determine the emotional atmosphere of the interrogation; the inflections of his voice will influence the subject's physiological reactions to the questions. And the polygrapher, in most real-life situations, will also score the polygraph charts, a subtle process in which the examiner's predispositions can easily determine marginal decisions. Psychologists know of a large body of research on something called the experimenter expectancy effect, of which a typical study might involve recruiting graduate students as experimenters and then telling them (erroneously) that theory or previous findings indicate that the subjects in this experiment will behave one way rather than another.[16] Whether the experimental subjects are humans or white rats, such experiments tend to produce the results that the student-experimenters (the actual subjects of the true experiment) have been led to expect. Sometimes, no doubt, these biased outcomes are a result of actual cheating; the student-experimenters fake the data so that the senior researcher will be pleased with their work. But this bias will occur without any deliberate falsification. Expecting (and hoping for) a certain outcome, the student-experimenters unwittingly influence the subjects' behavior in the expected direction. Or they recheck findings that come out "wrong" but not those that come out "right"—so that errors in the desired direction are selectively retained. Because of this research, most psychologists would agree with the following generalization: If a test procedure is sufficiently subjective and unstructured so that an unethical examiner *could* easily bias the results, then it is likely that even an ethical examiner, expecting or hoping for one outcome rather than another, *will* bias the results without any conscious intention to do so.

These considerations are especially relevant to the proposal that criminal defendants be permitted to offer evidence of having passed polygraph examinations. Such a practice unquestionably generates an active market for "friendly" polygraphers—and one must doubt that the interests of justice would be well served by the result. David Raskin is said to have charged his clients fees of up to $40,000, a reward calculated to induce a very friendly attitude indeed on the part of the examiner. As we saw in the DeLorean case, Raskin and his polygraph-computer hybrid found DeLorean to be truthful beyond doubt, whereas the subsequent FBI test found him to be unquestionably deceptive in denying his guilt.

The Polygrapher as Expert Witness

Professor Edgar Jones cites a query by an appellate court justice that poses the theme for this section; this justice wondered "why there was any difference between the testimony of a polygraph operator who believes the defendant is telling the truth, and of a physician who gives his expert opinion on a medical condition."[17] I was asked a similar question by a judge presiding at a murder trial in Anchorage, Alaska, in 1977. I had not thought that particular matter through before and my extemporaneous reply was halting and unsatisfactory. Here is what I should have said.

The law recognizes the special category of expert witness because courts often require information that is available only in the form of the opinion of some specially trained and knowledgeable person. Juries and judges are not competent to give medical tests or to interpret the results, to take fingerprints and then match them, or to determine whether a defendant is afflicted with a mental illness. Therefore, while ordinary witnesses are restricted to reporting what they did or saw or heard, a qualified expert witness is permitted to offer his opinion, based on his special fund of knowledge and experience. These opinions are inevitably fallible but they provide items of evidence that would be wholly inaccessible to a lay jury without the expert's help, pieces of the final puzzle that it is the jury's job to assemble into the most coherent picture that it can. But the polygrapher's evidence, also fallible, goes directly to the heart of the issue; it is offered, not as a piece of the puzzle but as the final picture complete. If the defendant was truthful on the lie test, then he is innocent; if he was deceptive, then he is guilty. Admittedly, there are cases in which another kind of expert's testimony may be crucial; if the fingerprints on the gun are those of the defendant, then, given the other facts in evidence, logic may compel the jury to bring in a verdict of guilty. But, even here, it is the jury's job to draw the necessary inference. When, on the other hand, a polygrapher asserts that, in denying his guilt on the polygraph test, the defendant was deceptive, there is nothing left for the jury to do but rubber-stamp this predetermined verdict or else to reject the polygrapher's testimony altogether.

Moreover, judges and juries *are* quite capable of assessing the credibility of a witness without any help from a polygraph examiner. Indeed, assessing credibility of witnesses is traditionally one of the main responsibilities of triers of fact. What I have called "clinical examiners" of the Keeler–Reid–Arther tradition use much the same sort of evidence in deciding credibility that courts customarily employ, including "behavior symptoms" and evidentiary considerations, and, sometimes, polygraphers use the sorts of unverifiable or hearsay evidence that our courts specifically reject. It is unlikely that any court would permit some Certified Public Logician, as an expert witness, to tell the jury how it ought to *combine* the facts in evidence in arriving at its verdict; making rational inferences from the facts at hand is another responsibility traditionally reserved to judge and jury. Accrediting as an expert witness a clinical polygraph examiner is an equally radical break with the basic principles of our system of justice. To take such a step merely because the parties have stipulated to it in advance makes as much sense as agreeing, because the parties have so stipulated, to settle the case through a trial-by-combat rather than by the verdict of a jury.

Admittedly, the average juror cannot administer a lie detector test or interpret a polygram. A polygraph examiner of the school that bases its conclusions entirely on the charts, eschewing "behavior symptoms" and other sources of information, could claim that his training allows him to assess credibility in a special way, not available to the court except through his opinion. Here again, it seems to me, the question of validity is central. If the polygraphic lie test were as nearly infallible as some of its advocates contend, then perhaps the polygrapher should be qualified as an expert witness (*the* expert, indeed, making all others redundant) even though his testimony goes to the heart of the issue and usurps the traditional function of the trier of fact. But the evidence suggests that the polygraphic lie test has only modest validity, that it is correct from perhaps 60% to 75% of the time. The base-rate considerations of the preceding section indicate that, for the courtroom applications now used or contemplated, this average validity may be reduced to chance levels of 50% or even lower. Instead of improving the jury's assessment of credibility, such evidence seems far more likely to pollute it.

The polygrapher clearly is a unique kind of expert witness, one whose presence in the courtroom constitutes an abrupt departure from traditional principles. For an institution as wedded as the law is to precedent and to the deliberate and prudent pacing of its own evolution, one would expect such radical steps to be taken only at the prodding of compelling evidence that these steps are in the right direction.

How Juries React to Lie Test Evidence

The Mendoza Murder Case   During the summer of 1974, James Ray Mendoza, age 19, was tried in a Wisconsin court on two counts of murder in the first degree. Two off-duty policemen, dressed in plain clothes, were shot on the street across from a Milwaukee barroom at about 2 A.M. Both fatal bullets were fired from a service revolver carried by one of the officers, who, postmortem tests indicated, had been drinking heavily.

Mendoza admitted the killings, pleading self-defense. He said that the men had accosted him after the bar closed, that they had beaten him, bending him back on the hood of a parked car and beating him with a gun butt. Eyewitnesses corroborated this account. Mendoza's hairs were found on the hood of the car. The defendant contended that in the melee he was able to seize the policeman's revolver with which he then shot both his assailants.

In view of the evidence, the prosecution offered to reduce the charges to second-degree murder and manslaughter if Mendoza passed a lie detector test. In return, he had to agree that the lie test evidence could be used against him if he was diagnosed to be deceptive. The prosecution's offer was accepted. Mendoza was tested by the chief polygrapher of the Wisconsin State Crime Laboratory. The examiner's verdict was "deceptive."

Over defense objections, the trial was moved from Milwaukee to the rich farm country of Sparta, Wisconsin. A non-Hispanic jury was impaneled. The defense brought in two other experts to dispute the validity of the lie test results but the trial judge interpreted Wisconsin case law to forbid disputing the polygrapher's findings through the testimony of rebuttal witnesses.

This jury was faced with an unusually clear choice. On the one hand, much of the physical and eyewitness evidence clearly supported the defendant's claim of self-defense. On the other hand, the polygrapher testified that a lie detector test had shown Mendoza's account to be deceptive. The jury chose to believe in the lie test. Mendoza was found guilty as charged. The judge, also apparently persuaded by the polygraph, sentenced the defendant to two terms of life imprisonment—to be served consecutively.

On appeal, the Wisconsin Supreme Court ruled that the proceedings should not have been arbitrarily moved to Sparta and ordered a new trial. Released on bail, James Ray Mendoza disappeared and was still a fugitive four years later when I last inquired. Who can blame him?

It is doubtful that even the prosecution believed that Mendoza had premeditated these killings, that he had planned to steal a gun from one of these plainclothesmen, shoot them both, beat himself with the gun butt, and then bribe several witnesses. As is generally true in such cases, the offer of dismissed or reduced charges in return for stipulation to the lie test suggests that the prosecution knew that the defendant's explanation might be persuasive without the polygrapher's expert opinion to refute it. By itself, the Mendoza case demonstrates that an American jury can turn its face from other evidence and allow itself to be guided by the verdict of the polygraph.

The Fay Murder Case   At 4:30 A.M. on March 29, 1978, a Conrail employee named Floyd "Buzz" Fay was awakened in his trailer home in Perrysburg, Ohio, by a loud banging on the door. Opening the door, Fay was confronted by several police officers, guns drawn, who promptly arrested him on a charge of aggravated murder. Fred Ery, manager of Andy's Carry-Out, an acquaintance of Fay's, had been shot in a robbery the night before by a man wearing a ski mask and carrying a sawed-off shotgun. Sedated for pain and having lost a lot of blood, the victim was asked who his assailant might have been. He replied that "It looked like Buzz but it couldn't have been." Five hours later and at the point of death, Ery seemed to have changed his mind and he said, "Buzz did it." But a search of the house turned up no shotgun, and no ski mask. Fay's only prior offense had been a ticket for driving "under the influence" five years earlier. A witness denied that the ski jacket found in Fay's trailer looked like the one the shooter was wearing.

After a week of unproductive investigation, the police offered to drop the charges if Fay would submit to a polygraph test, stipulating to its admissibility should he fail. Badly advised by his attorney, Fay agreed and failed the test. The district attorney proved to be unusually generous in keeping his side of the bargain. He brought in a second polygrapher from the respected firm of Lynn Marcy in Dearborn, Michigan, to test Fay again. Once again he was diagnosed as deceptive. At trial, no evidence was offered by the defense to refute the conclusions of the polygraph examiners. Fay was found guilty of aggravated murder and the one consolation his attorney could offer was that the usual penalty for this crime had been execution but the Supreme Court had declared Ohio's death penalty to be unconstitutional just a year or so before. Fay was sentenced instead to life in prison.

I first made Fay's acquaintance when he wrote to me, from prison, asking for reprints of some of my articles relating to the lie detector. He had retained the services of a bright young public defender, Adrian Cimerman, and I assured him that I would be willing to testify about the invalidity of the polygraph if Cimerman were to be successful in obtaining a new trial. But Mr. Cimerman did better than that. Like a true-life Perry Mason, he investigated the case and helped to obtain a confession from the man who had served as the driver of the getaway car, who, in turn, identified the shooter and the lookout. Fay was finally exonerated and, after serving two years of his life sentence, walked away a free man (Figure 18.1).[18]

There are other clues as to the way in which juries may react to the testimony of polygraphers. Following the first reported case in which lie test evidence was admitted, jurors were questioned by mail as to what importance they had attached to the polygraph findings. Six of the ten jurors who responded said that they considered the lie test results to be conclusive.[19] In another study, 20 third-year law students were asked to decide an imaginary case; all 20 found the defendant not guilty. They were then asked to suppose that there had been testimony that the defendant had failed a lie detector test. When told that the lie test was 99.5% accurate (roughly the accuracy claimed by Reid and by Arther), 17 of the 20 changed their verdict to guilty.[20] A similar study in Canada employed an imaginary murder case in which the defendant was reported to have confessed his guilt.[21] In spite of psychiatric testimony that this defendant was unstable and prone to false confession, more than half of a group of 50 mock jurors found him guilty. The case was then presented to another group of mock jurors, this time including testimony that the defendant had passed a lie detector test. With this one change, 72% now considered him to be innocent. For a third group, a statement by the judge was added, asserting that the polygraph was only 80% accurate and that they should be cautious in evaluating that evidence. Nonetheless, 60% of this group rendered a verdict in accordance with the lie test.

The one remaining study has been interpreted to indicate that juries may not be overinfluenced by lie test evidence.[22] Thirty-one jurors who had participated in moot court or practice trials were contacted by mail after they had already rendered their decisions. They were asked whether they would have voted differently if the evidence had included the results of a lie detector test. Only 20% acknowledged that such testimony would now convince them that they had voted incorrectly. But these jurors had already settled on the opposite verdict after lengthy deliberation. It is obviously more difficult to get people to change a settled opinion than it is to influence that opinion while it is being formed.

The impact of lie detector evidence on a jury will doubtless vary with the nature of the case, the impression made by the polygrapher, the quality of the rebuttal testimony if any, and the way in which the judge instructs the jury to interpret such evidence. More and better research on this issue would be useful. But those who scoff at fears that lie test evidence might "somehow be so prejudicial in its weight and impact that the jury will disregard all other evidence and go on the polygraph test results alone"[23]

Conviction In Perrysburg Murder Called Mistake; Man Freed After 2 Years

New Evidence Links Slaying To 2 Others

By JIM YAVORCIK
Blade Staff Writer

A former Perrysburg Township man, imprisoned more than two years for a murder which authorities now say he did not commit, walked out of the London Correctional Institute Thursday apparently a free man.

Meanwhile, Wood County prosecutor John Chestwood said newly discovered evidence in the murder case — which stemmed from the shooting death of a Perrysburg carry-out store owner — has led to the arrest of one suspect and a warrant being issued for the arrest of another.

Floyd Fay, 28, formerly of 10835 Ford Rd., was convicted of killing Fred Ezy, 24, inside Andy's Beverage Center, 134 East Third St., Perrysburg, by a Wood County Common Pleas Court jury in August, 1977. Mr. Ezy owned the store along with his father, Andy.

Fay Protested Innocence

Mr. Fay has protested his innocence since the day he was arrested.

Based on the new evidence, the prosecutor and Mr. Cimerman filed a joint motion for a new trial, which Judge Gale Williamson granted Thursday afternoon. The judge allowed Mr. Fay to be released without bond, pending final disposition of the case.

Since the new trial puts the case back to "square one," the prosecution can now dismiss the case by asking the judge for a dismissal in open court. Mr. Fay's next court appearance is scheduled for Nov. 17.

When the judge approves that request, the case of the State of Ohio vs. Floyd Fay will be history.

Mr. Chestwood noted that the criminal justice system generally works well, and that he was not aware of any case in which an innocent man had been convicted in the county.

He said Mr. Fay received a fair trial and a jury of 12 Wood County residents found him guilty based on evidence presented. But, "the new evidence compelled us to work as hard to see an innocent man freed as we'd worked to convict a guilty man," the prosecutor said.

Prosecutor Acted 'Diligently'

Mr. Cimerman, a county public defender just two years out of law school, said the prosecution acted "diligently" on their investigation after he pointed them with leads.

He said the case is a lesson to all that although our justice system is the best known to man, "it is not perfect."

He added that the failure of Mr. Fay to testify at his trial probably led the jury to its finding of guilty.

A criminal defendant has the right not to testify and that should not have been held against him, Mr. Cimerman said. The burden is on the prosecution to prove a man guilty, not on the accused to prove his innocence.

'I Made It,' Jubilant, Relieved Fay Says

MR. FAY, LEFT, TALKS WITH HIS ATTORNEY, MR. CIMERMAN
Pair discuss case as they leave prison after release
— Associated Press telephoto

Figure 18.1. Floyd Fay, exonerated by the confession of the real killers, is finally released from prison. © 1980 *The Toledo Blade*, used with permission.

will have a hard time convincing Floyd Fay—or James Ray Mendoza, if they can find him.

The Psychopathic Liar

Professor Edgar Jones points out that lie detection and truth verification are not quite the same thing, due to the limited capacity "of humans ... realistically to perceive, and then to store their perceptions as reliable recollections which can later be accurately summoned and recounted." To our capacity to misremember or to misperceive events there are many contributors, including "that ancient sixth sense—the intuitive leap of recognition that so often in fables alights on iridescent truth, while in life as often plunges into some mudhole of surmise."[24] If there could be a dependable lie detector, which Professor Jones rightly considers to be doubtful (it is not plausible that our species evolved through natural selection the equivalent of Pinocchio's nose), even then the lie detector's diagnosis could not assure us that what this witness says is true, but only that the witness believed it to be true at the time of testing. And most judges understand that there are people, with the best intentions, who are wholly unreliable as witnesses when their own interests and emotions are involved, because they so readily believe what they want to believe.

The psychopathic liar presents a very different problem because his unreliability is entirely conscious and deliberate. Psychiatrist Hervey Cleckley provides a characteristically vivid description:

> The psychopath shows a remarkable disregard for truth and is to be trusted no more in his accounts of the past than in his promises for the future or his statement of present conditions. He gives the impression that he is incapable of ever attaining realistic comprehension of an attitude in other people which causes them to value truth and cherish truthfulness in themselves. Typically, he is at ease and unpretentious in making a serious promise or in (falsely) exculpating himself from accusations, whether grave or trivial. His simplest statement in such matters carries special powers of conviction. Overemphasis, obvious glibness, and other traditional signs of the clever liar do not usually show in his words or in his manner. Whether there is a reasonable chance for him to get away with the fraud or whether certain and easily foreseen detection is at hand, he is apparently unperturbed and does the same impressive job.... During the most solemn perjuries he has no difficulty at all in looking anyone tranquilly in the eyes.[25]

The psychopath's talents as a liar are not due merely to long practice or the cultivation of a beguiling poker face. Research has shown that the genuine psychopath (the term is carelessly used to include people unlike Cleckley's prototype) is less bothered than the normal person is about danger in general and punishment in particular. He is relatively untroubled by impending unpleasantness, whether of his own making or not. A classic laboratory demonstration of this peculiarity involves connecting the experimental subject to a polygraph (which, while it cannot detect lying, can detect emotional disturbance) and then having him listen to a countdown: "ten ... nine ... eight ...," knowing that a tremendous blast of noise or a painful electric shock will be presented when the countdown reaches "zero!"[26] The polygraph shows the normal individual's apprehension beginning early in the countdown, increasing to a climax of high arousal just before the noxious stimulus is due to be delivered. The real psychopath, in contrast, shows relatively little premonitory perturbation. He does not look very worried externally and the polygraph confirms that his apparent insouciance is more than skin deep.

The simplest theory to account for these and other characteristics of the psychopath is that he has a low "anxiety IQ" (there is nothing at all wrong with the psychopath's actual IQ, his intelligence). Like every other biological characteristic, the innate tendency for fearfulness varies from one person to another. People at the high end of this continuum are easily frightened, timid, predisposed to certain types of neurotic disorder. Those at the low end would appear to be blessed; who needs worry and anxiety? But fear, like pain, has an adaptive function, most obvious in the formative years, and our society depends heavily on the sequence of punishment-fear-avoidance for the socialization of children. A child with a low anxiety IQ is less influenced by punishment. If his adult community responds by increasing punishment and disapproval, the underanxious child merely becomes alienated from the community and seeks his gratification elsewhere. And the lack of normal timidity and shyness does have advantages; the latent psychopathic child tends to be popular with his peer group; he is successfully aggressive, a leader in the gang, learns how to manipulate others, how to get the things he wants through guile or predation. The low-anxiety child responds normally to positive rewards, to success and achievement and the admiration of others. If blessed with insightful parents who know enough to emphasize the positive, he can become socialized through ties of affection and the self-discipline of pride and turn out to be a hero or community leader. Otherwise he may remain unsocialized, a feral creature—and an unusually skillful and confident liar.

We have seen that the normal individual tends to "fail" the standard lie test because he is aroused (actually frightened, worried) by the relevant questions and the polygraph reveals this apprehension. It is natural to suppose that a psychopath would be less aroused by the relevant questions and therefore less likely to be classified as deceptive—whether he is lying or not. In a properly scored CQT, the only way to "pass"—to be classified as definitely truthful—is to respond more strongly to the so-called control questions than to the relevant questions. There is no reason to expect the psychopath to show this pattern. Instead, one might expect the psychopath to show minimal reactions to both the relevant and the control questions and, thus, to produce inconclusive results. But since the CQT is predicated on the respondent's expected fear of discovery, if he is guilty and lying, and since the psychopath's cardinal characteristic is his attenuated fearfulness, his lack of normal guilt or shame or concern about the consequences of his actions, then it is natural to suppose that the standard lie test should be relatively unsuccessful in detecting lying in the psychopath.

A study of this very question has led polygraphers to think that this expectation has been refuted.[27] Inmates of a prison in British Columbia were classified as psychopathic or nonpsychopathic and then invited to participate in the experiment. Half of the participants were asked to commit a mock crime, to "steal" a $20 bill, and then to take a lie detector test while pretending to be innocent. The other group was actually innocent of the "theft" and took the same test. All subjects were offered a prize of $20 if their lie test indicated that they were truthful in denying the theft. In the event, nearly all of the "guilty" subjects were scored as deceptive and nearly all of the "innocent" subjects were scored as truthful—96% accuracy overall—and the psychopaths were as successfully classified as were the nonpsychopathic inmates. This is truly a remarkable result, especially since these statistics are so much better than this same polygrapher, Dr. David Raskin, achieved when he scored the CQT charts of real criminal suspects in the earlier study by Barland and Raskin, where his average accuracy was only 72%. Raskin (sometimes) acknowledges that the psychological differences between laboratory and field situations are "profound":[28] Accuracies obtained in laboratory experiments like this prison study cannot be used to estimate the accuracy to be expected in real-life uses of the lie test. Moreover, Raskin himself both administered the CQT examinations in the prison study and scored them. If we picture those 44 inmate participants, each trying to win his $20 and half of them hoping to bring back to their colleagues the news that they had "beaten the lie detector," it is easy to imagine that a least a few of them made some revealing comment or exhibited a telltale expression, perhaps after the formal test was over and they were being disconnected from the polygraph. For example, if a "guilty" subject asked before leaving the experimental room, "Well, how did I do? Did I beat it?" then, when Dr. Raskin sat down to score the polygrams, it would be very difficult for him to ignore this extrapolygraphic clue. When I was asked to comment on the report of this experiment by the editor of the journal to which it had been submitted, I urged that Dr. Raskin have the polygrams rescored by another polygrapher who had no extraneous cues as to which subjects were which. Dr. Raskin did not do this, however, so we shall never know how the real accuracy of the CQT on this occasion compared with that achieved in the Barland and Raskin study.

If we cannot take seriously the results of this experiment as an estimate of lie test accuracy in real life, can we at least learn something from the fact that the psychopaths tested performed no differently than the nonpsychopaths? However well the polygraph can do at detecting lying, does it do equally well with psychopathic subjects? We had predicted that psychopaths should be less disturbed and apprehensive than nonpsychopaths when being tested as criminal suspects, less fearful or remorseful while lying in response to the relevant questions. But, in the gamelike context of this British Columbian experiment, there was no occasion for any of the subjects to be frightened or remorseful. The arousal responsible for the polygraphic reactions observed on these tests should not have been generated by such negative emotions but rather by the inmates' interest in the project, the challenge of the task, the desire to "beat the lie detector" and to win the $20. And there is no reason to expect the psychopath to be any less challenged or any less interested in winning: "When the felon's not engaged in his employment, or pursuing his felonious little plans, his capacity for innocent enjoyment is just as great as any honest man's."[29] This is, of course, precisely why the laboratory lie test, in which the subjects are motivated by curiosity or the offer of prizes, is a poor simulation of the real-life situation in which the subject is motivated by fear or guilt.

Dr. Raskin conducted the experiment we have been analyzing while the guest of Professor Robert Hare at the University of British Columbia.

Dr. Hare is one of the leading authorities on the psychopathic personality. After considering the criticisms evoked by the report of this experiment, Dr. Hare concluded that the original thesis—that the psychopath may be better able than the normal subject to "beat" a lie detector test administered in the course of real-life criminal investigation—remains to be disproven.[30] We can only agree.

The Polygraph in Criminal Investigation

Nothing we have said so far militates against the use of polygraphic interrogation as an aid in criminal investigation under police auspices. If lie test results are not as accurate as some have claimed, they may still be useful in setting investigative priorities. If Pat passes the lie test while Mike shows a classic deceptive result, focus your inquiry first on Mike's comings and goings—but don't let Pat leave town. With the polygraph in action, as many as 25% of crimes may be solved neatly and efficiently by means of a confession—but make sure that the confession can be verified. There is reason to believe that the Guilty Knowledge Test could be of great value in police work, only in special cases to be sure, but possibly in many of those important and difficult cases where help is most needed.

The great problem in police polygraphy is in restraining the enthusiasm and credulity that these methods inspire in many police officers. A case in the news as I wrote the first edition of this book concerned a priest accused of a series of robberies, who was cleared by the confession of another man who resembled the priest in both appearance and manner. But the priest had "failed" a lie detector test and the press quoted the local police as still convinced that "we had the right man the first time." The lie test is seductive because it is so clear-cut, no shadings of gray, no equivocating probabilities; the suspect either passes or he fails. It must seem a relief from the many, often dangerous, uncertainties that characterize police work. But if police polygraphy is not to be abused, then police officers must learn to accept that the lie test is never certain either, that some villains pass, that many who fail will turn out to be innocent, that not even a confession should be accepted without verification.

Whether it would be in the interests of justice to present to a jury the results of a Guilty Knowledge Test is an intriguing question that will be taken up in Part IV. The considerations of the present chapter, in my view, combine to make a strong case against *any* use of lie detector test findings in the courtroom. The lie test has modest validity at best and is biased against the truthful suspect—unless the test is administered by a "friendly" polygrapher, in which case the bias is merely reversed. The test's validity will tend to be even lower on the selected set of cases that would be offered into evidence, due to the base-rate phenomenon. The lie detector usurps from the trier of fact the central responsibility for determining credibility. If the lie test examiner is accredited as an expert witness, no other witnesses, expert or not, need apply. As we shall see in Chapter 19, it is not difficult to teach most guilty suspects how to "beat" the standard lie test, and it seems probable that psychopaths and experienced criminals can often accomplish this without special training. If it has any legitimate role in the criminal justice system at all, the lie detector should be used exclusively— and circumspectly—in criminal investigation, aiding the search for the sorts of hard evidence traditionally admissible in court.

Chapter 19

HOW TO BEAT
THE LIE DETECTOR

This detestable machine, the polygraph (the etymology of which shows that the word means "to write much," which is about all that can be said for it).... It is such an American device, such a perfect example of our blind belief in "scientism" and the efficacy of gadgets; and ... so American in the way it produces its benign but ruthless coercion.

—WILLIAM STYRON[1]

A deceptive subject might try to beat the lie test by inhibiting his physiological reactions to the relevant questions. Some people can attenuate their responses even to very strong or painful stimuli if they know when the stimulus is coming.[2] Because the pattern of a control question lie test is fixed, a sophisticated subject should be able to tell when the relevant questions are about to be presented. Some persons have much better control of their reactions than others do. There are even ethnic differences. When Bedouin tribesmen of the Negev desert were examined on the polygraph, they were found to be far less reactive than Israeli Jews, whether of Near Eastern or European origin.[3] Moreover, most people will become habituated to any stimulus, such as a question, that has been frequently repeated, reacting less strongly to that stimulus than they did at first. A criminal suspect who has been extensively interrogated might, as a result of this habituating repetition, become less reactive to the relevant questions on a lie test administered later.

273

During the 1960s, my university accepted a secret research contract from the Air Force to study the effectiveness of countermeasures against lie detection. My job in this project was to train the experimental subjects. They practiced controlling their responses to my questions while observing their own reactions on the polygraph—the technique now known as "biofeedback." When I thought they were prepared, I would send them on to the chief of our university police department, a polygraph examiner of long experience, who would administer a formal lie test. This work had just gotten well under way when a new university president canceled all secret research contracts, including ours. (I never understood why the Air Force insisted on the "secret" classification, since the only thing about our project that could really be kept secret was the source of the funding.) But we had gone far enough by then to convince me that some people could learn to attenuate their relevant responses and beat the lie detector in that fashion—but that it is very difficult for most people and probably impossible for many.

Methods of Beating the Lie Detector

A much more effective method of beating the lie detector, however, is to augment one's reactions to the control questions.[4] However disturbed one may be by the relevant questions, the scoring rules require that the examiner cannot diagnose "deceptive" if the control reactions are just about as strong or even stronger. Knowing the principles of the method, a subject can identify the control questions when the examiner goes over the list in the pretest interview. During the test, the subject will try to sit calmly, breathing regularly, while listening to and answering the relevant questions. After each control question has been presented and answered, he will do something to augment his response. Any self-stimulation that is not visible to the examiner will tend to increase the normal polygraph reaction, covert actions such as biting one's tongue or lip, tightening one's anal sphincter, or strongly contracting one's toes, for example. A tack secreted in one's sock can be used to produce a good reaction on the polygraph.[5] So too can psychological self-stimulation, such as trying to mentally subtract 7s *seriatim* from 924 as quickly as possible after answering "No" to the control questions. Tensing the arm or stomach muscles, by contrast, will produce give-away movements of the polygraph pens and must therefore be avoided.

Not knowing how to go about it, most unsophisticated subjects make no real effort to beat the lie detector or, when they do try, their efforts are easily detected. They cough or hold their breath or move in their chair or tighten their arm muscles under the blood pressure cuff. And these activities usually occur during or just after the relevant questions and, therefore, tend to augment the very responses that will lead to a "deceptive" diagnosis. Even if he is expecting a more sophisticated attempt at "beating the machine," the typical polygrapher is likely to be deceived by the approach outlined above. John Reid once denied my contention that criminals could beat the polygraph by self-induced reactions, which, he said, "are so obvious and unnatural that they are a clear indication of guilt."[6] He apparently forgot that he proved my point himself years ago. Using muscular contraction or pressure, he found "that all the typical blood pressure responses of deception can be produced artificially at will" and that "the manner in which these blood pressure changes were effected was imperceptible to the operator."[7] Just as the polygrapher depends on the respondent's naïveté for the lie test to work in the first place—exaggerating the test's accuracy, deceiving him with the "stim test," misleading him about the function of the control questions—so too might the sophisticated subject capitalize on the fact that most examiners do not expect skillful countermeasures and, not looking for them, will not see them. In the field experiments discussed in Chapter 8, the CQT was quite successful in detecting lying although it did little better than chance in detecting truthful responding. If, unbeknownst to the experimenters, guilty suspects in these studies had attempted skillfully to beat the lie detector by the methods discussed above, is it possible that the frequency of false-negative errors might also have approached—even exceeded—50%?

Methods Taught by Floyd Fay

I mentioned earlier my correspondence with Floyd Fay while he was serving the first two years of a life sentence for murder. His conviction, since proven to have been in error, resulted in large part from testimony by a polygrapher that this defendant had failed a stipulated lie test. At his request, I had sent Fay some information about polygraphic interrogation, including an article of my own that explains how one might attempt to "beat" the Control Question Test. After some months, I received from Fay a letter that read, in part, as follows:

> Since reading the article that you sent me ... I have been running my own experiment. The prison that I am in forces anyone that is suspected of violating a prison regulation into taking a polygraph. I have been able to get to nine of these people prior to their taking a test. Out of the nine that I KNOW were guilty of the "offense" that they were accused of, nine have beat the test! I realize that this is a small group to work with, but the 100% "hit rate" is nothing to laugh at. All I have done is have them read the article that you have sent me and then explain exactly what you were saying and they have all beat the test.

It would be difficult for a researcher to set up a controlled study to determine whether guilty suspects, to be tested under real-life conditions, could be trained to beat the lie test. Fay does not claim to be a scientist but I think he has helped to illuminate an inaccessible corner. As he remarks, nine out of nine is nothing to laugh at. Attorney F. Lee Bailey once offered a prize of $10,000 to "anyone who can beat the lie detector." I think that it would be only fair if Mr. Bailey would pay off this bet to Mr. Fay, in wholly inadequate compensation for Fay's two years spent in prison, falsely convicted by the lie detector that Bailey claims to be almost infallible.

Methods Taught by the Raskin Group

While he has his checkbook out, Mr. Bailey ought also to send some $10,000 presents to a number of students at the University of Utah. In laboratory research by Raskin and his colleagues,[8] "guilty" subjects were trained in the use of countermeasures to be applied while control questions were being presented during their examinations. The actual training consisted of instructing subjects

> to press their toes to the floor, to bite their tongue, or ... to count backward by 7s from a number larger than 200 when the control questions were asked. Each countermeasure subject was instructed to begin the countermeasure as soon as he or she recognized a control question, stop just long enough to answer, and then continue the countermeasure until the next question began. Each countermeasure subject was then read a set of questions from a typical CQT and was coached in using his or her countermeasure unobtrusively so that it would not be detected by the polygraph examiner during the subsequent test. None of the questions used in this practice test was used in the actual polygraph examinations, and subjects were not informed of the order of the questions during the examination. The countermeasure training required a maximum of 30 min.[9]

The countermeasure training did not require attaching subjects to a polygraph to give them the opportunity to learn how their countermeasure maneuvers affected their physiological recording. The results indicated that

> the mental and physical countermeasures were equally effective: Each enabled approximately 50% of the Ss to defeat [i.e., appear truthful on] the polygraph test.... Moreover, the countermeasures were difficult to detect either instrumentally [i.e., by inspecting the physiological records] or through observation.[10]

The Right Way to Beat the Polygraph

No good social purpose can be served by inventing ways of beating the lie detector or deceiving polygraphers. As Fay's prison example shows, the most avid students of such developments would be professional criminals rather than the innocent suspects and the truthful job applicants who now fall victim to the trust that we Americans invest in this technology. (On the other hand, if I were somehow forced to take a polygraph test in relation to some important matter, I would certainly use these proven countermeasures rather than rely on the truth and my innocence as safeguards; an innocent suspect has nearly a 50:50 chance of failing a CQT administered under adversarial circumstances, and those odds are considerably worse than those involved in Russian roulette.)

In the preceding chapters, I have tried to show that no test based on the polygraph can distinguish truthful from deceptive responding with high validity, and that it is unlikely that a real lie detector will ever be invented. Similarly, I have tried to show that there are no behavioral cues that an experienced observer could employ so as to become a "human lie detector" of such accuracy that courts, employers, and the rest of us should defer to his expert judgment. We have seen that polygraph tests and honesty questionnaires are being increasingly used in the United States to decide which job applicants should be hired and which employees should keep the jobs they have. And we have noted that this trend has continued in the absence of any evidence at all that either test actually differentiates between the trustworthy and the dishonest. We have discovered that American businesses, like the ancient nobility, maintain their own private police, armed with lie detectors rather than with clubs, and that employees suspected of crimes against the company are tried by polygraph and punished by dismissal, without the protection of due process.

Turning to the official criminal justice system, we have discovered that the lie test is finding its way more frequently into the courtroom. We have acknowledged that, if the lie test were as accurate as its proponents claim, then in the interests of justice we should not only admit lie test evidence at trial but we should base the trial verdict directly on the lie test findings. But, perhaps thankfully, we have noted that in fact the lie test is not nearly so accurate as its advocates contend, that its validity is likely to be even lower on the selected cases in which it is now admitted by some courts, and that the consideration of such evidence is certain to impair, rather than facilitate, findings of fact in the traditional way. The use of the lie detector by the police as an investigative tool, while subject to abuse like any other tool, is not inherently objectionable and may improve the efficiency of the administration of justice through its effectiveness in inducing confessions. But we have learned that such confessions, like those that were induced by the rack and thumbscrews, may not be valid and must always be carefully checked for authenticity.

For all these reasons, then, it seems apparent that the truth technology must be regarded as a growing menace in American life, a trend to be resisted and, it may be hoped, beaten by measures more dependable and lasting than secreting a tack in one's sock. The 1988 Employee Polygraph Protection Act was an important step in the right direction. Extending that act to cover federal, state, and local government employees should be the next step. I believe, however, that the only safe solution, the only way to truly beat the lie detector, is to demythologize it. If lawyers, employers, judges, legislators, and government bureaucrats knew what you know now about the lie test, then the menace would be manageable. The first purpose of this book is to contribute to that end.

Some polygraphers are merely greedy opportunists. One would not regret their having to move into some more useful line of work. But most polygraphers are honorable people, firmly convinced that they are building a respectable profession that will benefit society. One must regret having to turn one's face against this group, advocating reforms that would put them out of business. I agree that, if they could distinguish truth from falsehood with great accuracy, then professional polygraphers would eventually sit at every crossroad of American life, passing the virtuous and forestalling the malefactors. I accept the sincerity of those who believe that their judgments are accurate 95% or 99% of the time—but I know that they are wrong. These claims are wildly implausible and the available evidence denies them. This one critique will not dismantle a deeply entrenched mythology that American children come to accept along with Santa Claus and the Easter Bunny but that, unlike those harmless fairy tales, they continue to believe in as adults. At least I hope to shift the burden of proof, which polygraphers have always shirked, back where it belongs—onto the shoulders of the "truth" merchants themselves.

Part IV

DETECTING GUILTY KNOWLEDGE

*And the whole secret, power, and knowledge of their own
discovery is locked within them—they know it, feel it, have the
whole thing in them.*

—THOMAS WOLFE, *The Web and the Rock*, 1939

The types of polygraphic interrogation we have been concerned with so far have all been designed for lie detection. We have seen that the validity of these various methods is generally unproven and that the limited evidence available indicates that this validity is modest at best. I have explained why I believe that no amount of future research and development is likely to improve much on the present, rather bleak, situation. The prospects may be better for another and fundamentally different method of polygraphic interrogation that is intended to detect, not lying, but the presence of guilty knowledge. In Chapter 20, I describe my own initial experiences with the Guilty Knowledge Test, review the evidence so far available as to its accuracy, and explain why this method seems to me to have promise as a tool of criminal investigation. Because this method is not well understood even by many polygraphers, I illustrate its application in Chapter 21, together with the theory of the Guilty Knowledge Test, and the possibilities for its future development.

281

Chapter 20

ORIGINS OF THE GUILTY KNOWLEDGE TEST

The real use of [psychophysiological measurements] is therefore probably confined to those cases in which it is to be found out whether a suspected person knows anything about a certain place or man or thing.

—Professor Hügo Munsterberg, 1908[1]

In 1958 I agreed to supervise two freshman medical students who had been awarded summer fellowships. Bright and full of energy, they made short shrift of the project with which I had thought to keep them busy for three months. I needed a new experiment to last us through July and August. Intrigued by the polygraphic equipment in my laboratory, my two assistants had asked if I did any lie detector work, and I had been forced to admit that I knew nothing about the subject. Equipped as we were with time, facilities, and ignorance, we resolved to do an experiment on lie detection.

The Experiment

Using student volunteers as our experimental subjects, we would have them enact mock crimes and then try to separate the "guilty" from the "innocent" by means of their physiological reactions during a stan-

dardized interrogation. For simplicity, we elected to use just one physiological measure, the electrodermal response (EDR; also known as the
galvanic skin response, or GSR), a wavelike change in the electrical resistance of the palms and soles, associated with imperceptible sweating in
those regions. The EDR is an exceedingly sensitive indicator; almost any
stimulus will elicit the response, and as a general rule, the stronger the
stimulus—the greater its arousal value for the subject—the larger the
resulting EDR will be.

We decided to use two mock crimes, a "theft" and a "murder." Our
student subjects were assigned at random to one of four groups. A subject
in the theft group, for example, was met at the lab by Assistant A and given
the following instructions: He was to lurk outside a certain office in the
building until the occupant departed. Then he would slip into the office
and go through the drawers of the desk, seeking the object specified for
him to steal. He would remove this prize and hide it in an empty locker in
the corridor. A subject assigned to the murder group went to a different
office where he played a hand of poker with the occupant, losing heavily.
He then pretended to kill this opponent, using a weapon provided by
Assistant A. After the victim had enacted his death throes, the student
"murderer" was to hide the weapon in a predetermined place and then
flee. A third group was assigned both of these crimes, hence these subjects
were doubly guilty. Those in Group 4 were merely told to imagine that
they had been picked up for questioning with regard to two matters that
they knew nothing about. After completing this first phase with Assistant
A, each subject was then taken to the polygraph room and handed over to
Assistant B, who, of course, was kept ignorant of the group to which the
subject belonged. B's job was to conduct the interrogation and to determine from the EDR reactions whether the subject was guilty of theft or of
murder, or guilty of both or of neither.

When it came to designing the actual interrogation, my ignorance of
the lie detector field proved invaluable. I assumed then that "lie detector"
was mere journalistic sensationalism. For the commonsense reasons discussed earlier in this book, I believed that actual lie detection was probably
impossible. But we were interested (as are the police) in guilt rather than in
lying; Assistant B's task was to distinguish the guilty from the innocent.
And the important psychological difference between the guilty suspect
and one who is innocent is that one was present at the scene of the crime;
he knows what happened there; his mind contains images that are not
available to an innocent person.

But how could we detect such guilty knowledge? Because of this knowledge, the guilty suspect will recognize people, objects, and events associated with the crime. There is no way to observe recognition directly, no distinctive "recognition response" just as there is no specific "lie response." But when we show a murderer a photograph of his victim, his recognition will stimulate and arouse him in a way that a picture of a stranger would not normally do. If we show him five pictures in sequence, including the one of his victim, then we might expect to see one large physiological reaction and four smaller ones. To the innocent suspect, all five persons pictured are equally strangers and he has no reason to react differently to the "correct" alternative than to any other.

Suppose you were a subject in Group 1 of our experiment, having enacted the mock theft. During the interrogation, Assistant B asks you the following:

> The thief hid what he stole. Where did he hide it? Just sit quietly and repeat the names of these places as I say them. Where did the thief hide what he stole? Was it … In the bathroom?… On the coat rack?… In the locker?… On the windowsill?… In the office?

You will probably show an electrodermal response to each of these five alternatives even though this is a make-believe situation and nothing important is at stake. But you know that the thief hid the stolen billfold in the locker, because you are the thief and you remember doing it. Therefore, your EDR to the alternative "In the locker?" will likely be larger than your responses to the other alternatives. If you were innocent of the theft, the five locations might seem equally plausible and only chance would determine whether your EDR to "locker?" would turn out to be the largest or smallest or somewhere in between.

In our experiment there were six multiple-choice questions for each of the two crimes. Some questions related directly to the criminal act and others referred to incidental matters. For example:

> If you are the murderer, you will know that there was an unusual object present in the office where the murder took place. Just repeat the name of the object. Was it … A phonograph record?… An artist's easel?… A candy box?… A chess set?… A baseball glove?

For half of the questions, Assistant B did not know himself which alternative was correct until after he had finished the test; sometimes there was an easel in the murder room, sometimes a candy box or a baseball glove. We scored the tests by awarding two points if the EDR produced by the correct alternative was the largest, one point if it was the second-largest of the five associated with each question. Thus, with a six-item test, the guiltiest score would be 12 and the most innocent-appearing score would be zero.

The Results

The experiment was a great success. We had decided to classify as innocent anyone scoring 6 or less; 48 innocent suspects were tested and all of them scored in this range. Of the 50 guilty suspects tested, 44 received scores of 7 or higher and were correctly classified as guilty. One of our volunteers had been a resistance courier during the 1956 rebellion in Hungary, twice caught and intensively interrogated by the Soviet Secret Police. On both occasions he had successfully maintained his cover and had been released. This man happened to be assigned to our Group 3, guilty of both crimes, and he was correctly classified by Assistant B without any need for beatings or electrical stimulation of his soft tissues. He concluded that he had defected to a technologically superior society. Using our objective scoring system and only the briefest of interrogations, we correctly identified who was guilty (or innocent) of which crime with 94% accuracy. Not one innocent subject was misclassified.[2]

I began reading the polygraphy literature at this point in preparation for writing up this experiment for publication. It turned out that our basic idea had been familiar to polygraphers since the early days, embodied in the method known as the Peak of Tension Test, which is rather like a single item of a Guilty Knowledge Test. Reid and Inbau describe a number of instances in which this method was used to encourage a confession. One striking example concerns a presser who had stolen a substantial sum of cash found in the pockets of a customer's trousers. As the polygrapher named various sums, this subject showed a peak of tension at "$765?" He was so impressed by the examiner's ability to pinpoint the exact amount that he confessed and made restitution.

On the other hand, a one-item multiple-choice test is not very useful when the respondent fails to confess. My reading revealed that polygraphic interrogation in practice consisted almost exclusively of attempts at actual lie detection of the sort already described. I was persuaded that the Guilty Knowledge Test rested on sounder psychological principles (even if my claim to its invention was arguable) and that the GKT should be considerably more dependable than any method of lie detection especially for the exoneration of the innocent.

We quickly set up a second experiment. In this one, 20 sophisticated subjects—psychology graduate students, psychiatrists—were offered a money prize if they could "beat" the test, either by inhibiting their responses to the correct alternatives (which is difficult) or by producing augmented responses to incorrect alternatives through self-stimulation (which is easier). These subjects were attached to the EDR apparatus and given time before the test to practice whatever technique they had decided to use, watching the pen trace out their EDRs as they experimented. Since the prizes were to be paid out of my own pocket, I used 20 items on this test instead of only 6, and I also used a more elaborate scoring system. All 20 subjects were correctly classified by this improved version of the GKT.[3]

Replication by Other Investigators

When psychological experiments are repeated, by different investigators in different laboratories, all too often they do not come out the way they did the first time. There are many reasons for this, chief among them being that psychology is an inexact science attempting to comprehend the most difficult of subject matter. I was greatly relieved, therefore, when another researcher first repeated my guilty knowledge work and got similar results.[4] At least ten additional laboratory studies of the GKT have subsequently appeared; the method always works, but two of these later experiments did not obtain as good a discrimination as I did between "guilty" and "innocent" subjects.[5] A virtue of the GKT method is that, at least in theory, the discrimination of guilty from innocent suspects can be made as complete as one wishes, simply by increasing the number of good items. With ten good items, for example, one might expect to identify 99% of guilty suspects if persons scoring 6 or higher are classified as guilty. The odds against an innocent suspect scoring so high would be from 16 to 1, for 6 "hits" to more than a million to 1 for 10 hits. This suggests that the experiments that failed to achieve good discrimination either had too few items or else items that were not good enough.

Before explaining what is meant by a "good" item, let me define a few terms. The GKT items suggested for testing Demjanjuk, in Chapter 3, or O. J. Simpson, in Chapter 21, each have six alternatives. The first alternative in each set is called a "buffer"; it is incorrect and does not figure in the scoring. The purpose of the buffer is to dissipate the subject's tendency to react more strongly to the first item in any series. One of the other alternatives is correct, or relevant; the relevant alternative refers to the "key," the bit of guilty knowledge on which that item depends. The remaining four incorrect alternatives are called controls. Unlike the usage of this term in lie detection, the control alternatives in a GKT item are genuine controls in the sense that the responses they elicit provide an estimate of how strongly an innocent person should react to the relevant alternative.

There are several methods of scoring the GKT. The simplest is to award one point for each item on which the relevant alternative produced a larger response than did any of the controls. The respondent is then said to have "hit" on that item. With ten (good) items, the typical innocent suspect will hit on fewer than three items while more than half of the guilty suspects should hit on at least seven items. The *cutting score* is the highest score that will be classified as indicating innocence. For example, if the cutting score is set at 4, fewer than 1 innocent suspect in 100 would be expected to be erroneously classified as guilty. It will be recalled that this type of misclassification is called a *false-positive*. The *false-negative* rate in this example—guilty suspects misclassified as innocent—would be about 4%.

A Good GKT

Now that we have a working vocabulary, what do I mean by a "good" GKT item? An item with five scored alternatives will be "good" if (1) an innocent respondent has about one chance in five of hitting on that item, and (2) a guilty respondent has at least an 80% chance of hitting on that same item. The first condition will be realized if all five alternatives seem equally plausible to a person without guilty knowledge—and if an un-scored buffer is employed as the first alternative presented. The second condition, that most guilty suspects would hit on the item, will be realized if most guilty suspects unequivocally recognize the relevant alternative as part of their guilty knowledge. (This second condition also requires that the guilty suspect will not successfully self-stimulate on one or more of the control alternatives. I will return to the problem of countermeasures later.) Thus, the key bit of information on which the item is based must be something that the guilty person is likely to have noticed and remembered. The control alternatives should be easily distinguished from the relevant alternative (and from each other) so that the guilty suspect is not confused about which alternative he's seen before. For example, if the alternatives are photographs of six bank tellers and the relevant one pictures the teller that was robbed, the people chosen for these photographs should all be quite different in appearance so that the guilty suspect can know at once which faces are incorrect and which one he has seen before. This discrimination can be enhanced during the testing by requiring the subject to name each alternative in some fashion after it has been presented to him. If instead he merely repeats some useless answer like "No" or "I don't know" after each alternative, or if he is not required to vocalize at all, then it will be easier for a guilty suspect to think of other matters and thus minimize the difference in impact between the relevant and control stimuli. Finally, the fact that he recognizes the relevant alternative must seem important to the guilty suspect if the item is to work as a good item. This condition should always be realized in a criminal investigation context but may not be in some laboratory experiments.

Various GKT studies have reported accuracies ranging from about 70% to 100%. In the less successful studies, one can readily see how the accuracies might have been raised, either by improving the quality of items or by increasing the number of good items. In one experiment, for example, the guilty knowledge keys were six "code words"—like "blue, hat, window, day, air, north"—that all "guilty" subjects had memorized.[6] The purpose of the GKT was to determine who knew the code words. A typical test item would be, "Is one of these words a code word? Green. Yellow. Blue. Red. Brown. Pink." The subjects answered "No" after each alternative. There was no buffer alternative. Since all the alternatives in each item were of the same class—colors, articles of clothing, etc.—a "guilty" subject might respond to the first alternative because it is the first of a series; to the second alternative because it leads him to think that all the alternatives are going to be colors; to the third alternative because it is the key "blue"; to the fourth alternative because, expecting a color, he has predicted "red" and it is "red"; and so on.

If that experiment were repeated using a buffer alternative, randomly chosen words as controls, requiring the subjects to echo each word instead of answering "No," and with 10 or 16 such items rather than only 6, then one would have to predict much better accuracy than the 77% found by these investigators. The accuracy that can be expected from GKTs containing from 5 to 16 "good" items is shown in Table 20.1.

Most laboratory studies have yielded results that fit reasonably well with the predictions of Table 20.1. For example, Geison and Rollinson[7] required 20 guilty subjects to enact a mock crime and then tested them,

Table 20.1. Accuracy of the GKT for Tests of Increasing Length

# of items	Cutting score	Innocent subjects (%)		Guilty subjects (%)	
		True-negative	False-positive	True-positive	False-negative
5	3	99.4	0.6	87.2	12.8
6	3	98.3	1.7	90.1	9.9
10	5	99.4	0.6	96.8	3.3
12	6	99.6	0.4	98.1	1.9
16	8	99.9	0.1	99.3	0.7

Note. The table assumes all "good" items, i.e., that an innocent subject has a 20% chance— and a guilty subject about an 80% chance—of giving his largest response to the relevant alternative.

along with 20 innocents, using six well-constructed GKT items. According to Table 20.1, we should expect 98% of the innocent subjects to pass and 90% of the guilty to fail; the actual results were 100% and 95%, respectively.

Lieblich *et al.*,[8] using ten items, obtained 93% to 100% accuracy. Podlesny and Raskin[9] used five reasonably "good" items and correctly identified 100% and 90% of the innocent and guilty. The results of eight of the better laboratory studies of GKT validity are summarized in Table 20.2.

In discussing the studies of lie detector validity in Chapter 8, I argued that the results of laboratory experiments, in which volunteers are required to commit mock crimes and then lie about them during a polygraph test, cannot reasonably be extrapolated to the real-world situation of criminal investigation. The emotional impact of the relevant questions must be greater in real life when the subject has something important at stake, such as his job, or his reputation, or his freedom. At the same time, the impact of the so-called control questions becomes relatively less in the real-life situa-

Table 20.2. Aggregated Results of Eight Analog
Studies of the Accuracy of the GKT

Status of subject	Diagnosed "innocent"	Diagnosed "guilty"	Number of subjects	Percent hits
"Guilty"	19	142	161	88.2
"Innocent"	147	5	152	96.7
Totals	166	147	313	93

tion. The problem is that the "controls" are not true controls at all; they do not elicit a reaction from this subject in this situation that predicts the reaction he should give to the relevant questions if he is telling the truth. But the Guilty Knowledge Test has genuine controls. The incorrect alternatives elicit reactions that this subject in this situation should produce to the correct alternatives *if he does not know which alternative is correct*. In the gamelike situation of the laboratory, the emotional impact of all of the questions may be less than it would be in real life. If so, then the subject's responses to all of the alternatives will be smaller in the former situation than in the latter. But in the laboratory as well as in the police station, the subject *cannot* consistently respond more strongly to the correct than to the incorrect alternatives if he does not know which is which. Theoretically, the GKT should work as well in real life as it does in the laboratory—but that theory needs to be tested by experiment.

The ability of the GKT to detect both innocence and guilt decreases when there are fewer items and also when there are fewer alternatives per item. The validities that have been obtained in the laboratory studies of the GKT that have been published to date have been close to those predicted from the numbers of items and alternatives per item used in each study. In Table 20.1, the GKT's validity with innocent subjects was predicted on the assumption that, with five scored alternatives per item, an innocent person would have one chance in five of giving his largest response to the correct alternative. The GKT's validity with guilty suspects cannot be predicted with such confidence because we cannot be certain that each suspect will have noticed and remembered all ten of the items of guilty knowledge; in Table 20.1, I assumed an 80% chance of remembering each correct alternative. The only investigators to study the GKT in actual criminal interrogations, Elaad and his colleagues in Israel,[10] found that their guilty suspects averaged about a 70% chance of responding most strongly to the correct alternative, as compared with about 80% for subjects involved in the mock crimes of the laboratory studies where the details of the crime scene were still fresh in their minds. On that basis, as shown in Table 20.3, we can say that, for a suspect failing two items on a ten-item GKT, the odds are more than 200 to 1 that he is innocent. For a suspect who gives his strongest response to the correct alternative on nine of the ten items, we can say that the odds of his possessing guilty knowledge are greater than 30,000 to 1.

The Israeli field studies achieved 97% detection of innocent suspects but only 76% detection of guilty suspects, which has been cited as indicating a defect of the GKT, but this is an erroneous conclusion. Elaad *et al.*

Table 20.3. Odds of Guilt vs. Innocence
for Various Outcomes of a Ten-Item GKT

No. of hits	Percent innocent with this score	Percent guilty with this score	Odds of innocence versus guilt
0	10.74	.00059	18,203:1
1	26.84	.01378	1,948:1
2	30.20	.14467	209:1
3	20.13	0.5009	40:1
4	8.39	3.6757	2:1
5	2.64	10.2103	1:4
6	0.52	20.012	1:38
7	0.08	26.683	1:334
8	.007	23.347	1:3,335
9	.0004	12.106	1:302,656
10	.00001	2.8248	1:282,475

Note. Table assumes a ten-item GKT with five scorable alternatives per item, and also a probability of 70% that a subject with guilty knowledge will "hit" on any given item. If suspects scoring 4 or fewer are classified as innocent, while those scoring 5 or more are called guilty, the accuracy in clearing innocent suspects will be greater than 96% and the accuracy in identifying guilty suspects will be greater than 95%. If guilty suspects remember 80% of the correct alternatives, both accuracy rates rise to greater than 99%.

used GKTs with only one to six items (mean = 1.8), each repeated typically three times, so that their detection efficiencies were predictably less than would be expected with GKTs constructed from six to ten different guilty knowledge facts.

Beating the GKT

Like the CQT, the GKT may also be subject to defeat by guilty subjects who self-stimulate to augment their autonomic responses to the incorrect or control alternatives.[11] However, some physiological responses are easier for the subject to control than others. For example, "voice stress" devices, such as the CVSA discussed in Chapter 11, might work very well with the GKT. Although they cannot measure lying, if they measure anything consistently at all they should be able to show that the correct alternatives elicit some difference in response than the incorrect alternatives, but only if the subject possesses guilty knowledge. It is not at all obvious how one might learn to control the microtremors in one's voice so as to beat the

GKT. Moreover, several recent studies have demonstrated that high accuracy can be achieved with the GKT when brain potentials are recorded and used to determine the presence of guilty knowledge.[12] Because such potentials are derived from brain signals that occur only a few hundred milliseconds after the GKT alternatives are presented, and because as yet no one has shown that humans can alter these brain potentials at will, it is unlikely that countermeasures could be used successfully to defeat a GKT derived from the recording of cerebral signals.

Chapter 21

FORENSIC USES OF THE GUILTY KNOWLEDGE TEST

A little [knowledge] is a dangerous thing.

—ALEXANDER POPE

During the year prior to the infamous murders of Nicole Simpson and Ron Goldman in Los Angeles, my colleague Bill Iacono and I, together with then-Director of the Department of Defense's Polygraph Institute (DoDPI) Dr. William Yancey, traveled to that city to meet with certain members of the L.A. Police Department. These included the head of the LAPD's polygraph unit, the chief crime scene photographer, and several experienced detectives of the homicide and major crime units. The purpose of our trip was to explore the possibilities of conducting an exploratory study of the forensic applications of the guilty knowledge technique in cooperation with the LAPD. One or more local psychologists, trained by us, would carry a "beeper" by which he could be summoned by one of the participating detectives to any fresh crime scene that seemed promising—that is to say a crime scene that gave the appearance of posing what the police call a "Who done it?" problem. Our psychologist, with—we hoped—the increasingly enthusiastic cooperation of the detectives and criminalists, would seek and record facts and pictures of the scene that had promise as the basis for GKT items. The criminalist-photographer we talked to was at once interested and pointed out that present-day methods of digital photography make it possible to easily rearrange the elements in

a crime scene photograph, thus providing numerous "incorrect alternatives" that only a suspect who had witnessed the scene would be able to
distinguish from the original photograph.

Our plan, which Dr. Yancey was willing to underwrite with a grant
from DoDPI, was to begin just on a catch-as-catch-can basis, accumulating
practical experience, until such time as we had learned enough to set up a
controlled study. We also believed that once several successful cases had
been seen through to the end, the interest of the detectives and criminalists
would be great enough, and their understanding of the technique clear
enough, that we could dispense with the beepers and leave it to those
professionals to locate GKT "keys" on their own. Whenever possible,
persons identified as suspects in these crimes would be administered a
GKT as well as a conventional lie detector test.

Alas, after two very promising visits, we suddenly got word that
LAPD had decided not to participate after all. We never learned the reason
for this change of heart but my suspicion is that it was the polygraphers
who dug their heels in at the end. With the detectives and criminalists
doing the real brain work, there at the crime scenes, the role of the polygrapher would become that of a mere technician. In fact, the best practice
would be for the investigators not to divulge to the polygraphers which
answers to each question were correct. Only then could one be sure that
the correct alternative was not somehow communicated to the suspect
unconsciously by the examiner. I think the LAPD polygraphers are happy
in their present status and reluctant to change it.

In any case, how sad it seems in retrospect that our project was not in
high gear a few months later when a report came in of an apparent murder
of a young man and woman in the Brentwood section of Los Angeles.
What turned out to be the notorious O. J. Simpson murder case, which
greatly damaged the reputation of the LAPD and which many believe
made a mockery of American criminal justice, might have turned out very
differently indeed if the detectives, photographers, and other criminalists
first on the scene had known enough to clamp down an impenetrable, if
temporary, veil of silence about what they found there and what could be
inferred from those findings, and then to systematically identify as many
facts and visual impressions as possible that would be likely to exist within
the mind of the last living person on that scene before the police were
summoned, the murderer himself. This case serves as well as any to refute
the claims of critics that the GKT would not often be a useful forensic tool.
Had "our team" been in place that fateful night, I think that police investi-

gators the world over would be busily learning about the Guilty Knowledge Test—perhaps even buying this book for their department libraries.

Did O. J. Simpson Kill His Wife?

This highly publicized case illustrates once again how the Guilty Knowledge Test, skillfully employed, might contribute to the prompt and just resolution of even high-profile crimes. Critics of the GKT insist that it could seldom be used because the details of the crime become widely known so that they cannot be used in GKT items. Many of the items suggested below could have been used in spite of the publicity surrounding this case because they refer to facts that were not publicized since they seemed incidental or irrelevant. Moreover, it is hard to believe that other details could not have been kept out of the media if the police appreciated the importance of protecting such information for at least a few days.

Another objection to the GKT, which applies also to the conventional polygraph test, is that defense attorneys are certain to refuse to permit their clients to cooperate. In the Simpson case, at least at the outset, it seems possible that Simpson's lawyers believed that he was not guilty. Supposing their client innocent, they should have been much distressed by the realization that the entire resources of the LAPD were being devoted, not to solving the case, but to incriminating that particular accused. Therefore, I think they might have been willing to listen to the following argument:

> If Mr. Simpson betrays guilty knowledge on this test, we will continue to believe that he is guilty and to seek the evidence that will convict him; you and he would then be in the same position you are in right now. But, if the test shows that he does not have guilty knowledge, then we will deploy our resources searching for some other perpetrator now, before the trail gets cold. If your client is telling the truth about not having been at the crime scene, the GKT should show that. The best way for you to really clear your client, and protect him from a stressful and perilous experience, is to help in this way to convince us that we should be looking elsewhere for the murderer.

In the following list of illustrative GKT questions, I have included a few (marked with an asterisk) that assume that even basic crime details could have been kept secret for a few days. I have also made use of the fact that crime scene photographs can now be modified *ad libitum* by computer techniques, removing, adding, or rearranging objects in the scene.

A Guilty Knowledge Test for O. J. Simpson

* 1. You know that Nicole has been found murdered, Mr. Simpson. How was she killed? Was she drowned? Was she hit on the head with something? Was she shot? Was she beaten to death? Was she stabbed? Was she strangled?

* 2. Where did we find her body? Was it: In the living room? In the driveway? By the side gate? In the kitchen? In the bedroom? By the pool?

3. What was Nicole wearing when she was killed? Was she wearing pants and a shirt? Was she wearing shorts? Was she wearing a skirt? Was she in a bathing suit? Was she wearing a bathrobe?

4. What was Goldman wearing when he was killed? Was it: Swimming trunks? A dark-colored suit? A striped sport shirt? A light-colored suit? A white short-sleeved shirt? A dark sport coat?

5. Nicole was dressed in one main color when we found her. What color was she dressed in? Was it: White? Black? Pink? Blue? Green? Tan?

6. I'm going to show you six pictures of Nicole's body, O. J. One of them shows her just as we found her. In the other five pictures, her body has been transposed to other locations, places where we might have found her but did not. Which of these pictures shows Nicole where we found her? Is it: This one? This one?… etc.

7. Goldman was carrying something in his hand when he was killed, something we found near his body. What was he carrying? Was it: A shopping bag? A sealed white envelope? A brown leather key case? A wireless phone? A liquor bottle? A small black address book?

8. An object was kicked or knocked over not far from the bodies. The murderer may have tripped on it. I'm going to show you six pictures of objects that might have been knocked over—only one of them is the object we found. Which of these objects was knocked over? (A croquet set) (A flowering plant in a big vase) (A small tea table) (A set of golf clubs) (An aluminum lounge chair) (A garden bench)

9. We know that whoever did this left a bloody article of clothing at the scene. It was either a shirt, a glove, a baseball cap, a handkerchief, a mask, or a shoe. What bloody article of clothing did the

killer leave behind? Was it: A shirt? A glove? A cap? A hand-kerchief? A mask? A shoe?

10. Now I'm going to show you a picture of Goldman the way we found him along with five other pictures with his body in different locations. Which picture shows Goldman lying as the killer left him? Is it: This one? This one?... etc.

These questions would compose a ten-item Guilty Knowledge Test. If Simpson happens to give his strongest physiological response to the correct alternative in no more than two of these ten items, then we can conclude that the odds are more than 200 to 1 that he is innocent and we should start looking for another suspect. If he responds most strongly to the correct alternative in as many as seven of these items, then the odds are better than 330 to 1 that he is guilty (see Table 20.3). One of the most serious problems about the Simpson case as it actually played out was the black–white polarization of opinion that occurred. Suppose that Mr. Simpson had "hit" on two, one, or none of these items; the police would have looked for, and perhaps found, another suspect. But suppose that as an examiner who did not know the correct answers read the ten items, Simpson gave his strongest physiological response to the correct alternative on eight or more of them; the odds against an innocent man's doing this are more than 8,000 to 1. I think most reasonable African Americans would have reluctantly concluded that this revered role model was guilty after all.

These ten items were generated in just a few minutes' time without even visiting the crime scene. A forensic scientist experienced in this technique and who had been at the scene while it was still in a pristine state could easily generate more and probably better items. It seems reasonable to conclude that whether O. J. Simpson did or did not kill his wife could have been determined with high confidence using a Guilty Knowledge Test administered within hours after he was first in police custody.

Who Blew Up the Murrah Building?

At 9:02 A.M. on April 19, 1995, a rented Ryder truck, packed with two tons of ammonium nitrate, blew up in front of the Alfred P. Murrah Federal Office Building in Oklahoma City, destroying the building, killing 168 men,

women, and children and injuring hundreds more, the worst terrorist
atrocity in American history. Within just two hours, one likely suspect,
Timothy McVeigh, was in custody. Within 24 hours, investigators knew a
number of facts that would have also been known to the bomber(s). They
knew what kind of truck had been used. They knew the ingredients of the
explosive mixture. They knew that the truck had been rented at Elliott's
Body Shop in Junction City, Kansas, on April 17. They could easily have
photographed the truck rental establishment used, as well as five others
chosen to be easily distinguished from the real one so that the bomber
would be likely to recognize the one he used on sight. Similarly, they could
have photographed the truck rental clerk behind his counter, dressed as he
was on the fatal day, and photographed also five other persons, behind
different counters, so that an innocent suspect would have no clue as to
which was which. The bomber used a false name, "Robert Kling," and a
fake South Dakota driver's license while renting the truck; with suitable
alternatives, these facts would provide three GKT questions. The bomber
would have had to make a deposit; how much was it? How was it paid, in
cash? by credit card? what kind of card? Does the clerk recall any conversation with his customer? any unusual remark that the bomber would be
likely to remember? What did the bomber say he wanted the truck for?
When did he say he would return it? Ryder rents a variety of trucks of
different shapes and sizes. The bomber should easily recognize which type
he drove from a series of photographs.

I have not made a special study of the facts in this case and yet I can
readily imagine a minimum of ten keys from which ten GKT questions
could have been quickly fashioned, within 24 hours of McVeigh's arrest.
Would all of these facts have been already published in the *National
Enquirer*? Not if the FBI agents were aware of the importance of maintaining security, and of course, many of the keys would not appear to be of
interest even to tabloid readers. Would the suspect have been available for
testing? Here is a sticking point, admittedly. If I were a public defender, I
would let my client take a properly constructed GKT on the grounds that I
would not construe my job to be to protect a guilty person from having his
guilt revealed. But McVeigh's public defender, Stephen Jones, and his staff
of 21 lawyers have, as of this writing, spent more than $10 million of
taxpayers' money in their effort to keep their client out of jail, and I
suppose that having access to such deep pockets (your pockets) tends to
cloud one's moral judgments.

I would like to think that the courts would eventually agree that requiring a subject to participate in a properly constructed GKT would not be a violation of the Fifth Amendment, which, after all, was designed to protect a criminal suspect from trial by torture, not from having his guilt discovered. I should think that the courts might acknowledge that submitting to a GKT was equivalent to submitting to a blood test for DNA analysis or to a lineup for witness identification. But the point I want to emphasize here is that with reasonable police work, the guilt or innocence of Timothy McVeigh could have been established conclusively within 24 hours of his apprehension. It seems to me that this would have been a good thing.

Scoring the GKT

The problem with the simple scoring scheme used in the examples above (and in the "Ivan the Terrible" case discussed in Chapter 3) is that an item is scored as a miss unless the relevant response is larger than all the control responses. But plainly, if a subject always gave his second-largest response to the relevant alternative, this too would strongly suggest guilty knowledge. By chance or because of covert self-stimulation, one of the four control alternatives might produce the largest response, but the reaction to the relevant alternative would be consistently larger than average. A method of scoring the GKT that is sensitive to such possibilities is the method of *mean ranks*. One ranks the responses for each item and then one averages the ranks of the relevant alternatives. The probability that an innocent suspect would produce such an average rank score can be determined from a table like Table 21.1.

In special situations where it is expected that the subject may try to conceal his guilt by deliberately augmenting his control responses, the GKT can be scored by the method of *expected ranks*. This method requires a larger number of items or else several repetitions of the item list. Since the innocent suspect cannot distinguish the relevant from the control alternatives, one expects that the relevant responses of an innocent suspect will about equally often be ranked 5, 4, 3, 2, or 1. Any statistically reliable deviation from equal rank frequencies will suggest that this suspect could in fact discriminate the relevant alternative—and that he has guilty knowledge. Normally, one would expect ranks of 1 or 2 for the relevant responses of a guilty suspect. Should he manage to strongly augment every control response, then the relevant responses may all rank 4 or 5. Even if every

Table 21.1. Probabilities of Various
Mean-Rank Scores for an Innocent Subject
on a Ten-Item GKT

Mean rank	Probability	Cumulative probability
1.0	.0000001	.0000001
1.1	.000001	.0000011
1.2	.000006	.000007
1.3	.00002	.00003
1.4	.00007	.0001
1.5	.0002	.0003
1.6	.0005	.0008
1.7	.001	.0019
1.8	.002	.0042
1.9	.004	.008
2.0	.007	.016
2.1	.012	.03
2.2	.019	.05
2.3	.027	.08
2.4	.037	.12
2.5	.049	.17
2.6	.060	.23
2.7	.071	.30
2.8	.080	.37
2.9	.086	.46
3.0	.088	.55

Note. For example, there are about 16 chances in 1,000 that a subject without guilty knowledge would produce responses to the relevant alternatives having an average rank of 2.0 or lower.

relevant response ranks exactly 3, the average of the control responses, we would be able to infer guilty knowledge. The suspect may have self-stimulated on two control alternatives for each item; how else could such a consistent result be achieved unless there was recognition of the relevant alternatives?

But I am letting myself get carried away. A policeman, eager for a conviction, could easily insure that his suspect will get a failing score on the GKT whether he is guilty or innocent. He could make sure that the only plausible alternatives are also the correct alternatives. Or he could read out the test questions in such a way as to emphasize the correct alternatives by his tone or inflection. We must reluctantly assume that any investigative technique that can be abused, will be abused sometimes. That is why the

GKT should be routinely tested on persons known to be without guilty knowledge, and the questions be presented by someone ignorant of the correct answers and unable, even inadvertently, to cue the correct alternatives. These should be standard practices.

Countermeasures

The GKT will be susceptible to distortion by a guilty suspect who is skillful enough to covertly augment his reactions to the controls. As is true also in lie detection, the best defense against such countermeasures is an observant examiner. Since the GKT is to be used only in criminal investigation, as a guide to the police, it does not much matter that a suspect invalidates the test in this way as long as his efforts are noticed. In lie detection, an innocent suspect might augment his control responses because of a justified fear of failing the test even though innocent. Only a guilty suspect would have reason to try to beat the GKT, the cardinal virtue of which is a vanishingly small likelihood of false-positive errors. More importantly, an innocent suspect *could not* systematically self-stimulate on the controls, since he would not know which alternatives are controls and which are relevant. Therefore, the police investigator learns just as much from the fact that a suspect tries to defeat the test as he would learn from a score in the guilty range. In both instances, he will focus his investigation on this suspect, searching for admissible evidence against him.

Under special circumstances, it may be worthwhile to attempt to get a valid GKT score in spite of obvious attempts by the respondent to defeat the test. This will require perhaps five presentations of a ten-item set in order that the method of expected ranks, described earlier, can be employed. When this technique was tested experimentally, 20 subjects were allowed to practice self-stimulation to produce misleading responses, and they were offered a money prize if they could "beat" the test; none were successful.[1] Whether the method will work as well in real-life applications remains to be seen.

The GKT and the Polygraph

So far I have assumed that the GKT will be administered while the suspect is connected to a polygraph and that the only response variable used in scoring will be the electrodermal response, the EDR. The EDR is an extremely sensitive indicator of momentary arousal, but as we have seen, the EDR is also relatively easy to augment through self-stimulation. I mentioned in Chapter 20 the possibility of using voice stress or brain wave responses because they would be hard for the subject to control. At least two other response variables might be used to supplement the EDR. One is the transitory change in heart rate that may follow each stimulus. This often takes the form of a brief acceleration, then a slowing, of the heartbeat. Using heart rate, then, one could consider the *pattern* of the response as well as its amplitude. There is no reason why the pattern or the size of the relevant and control responses should differ if the subject does not know which alternative is correct.

Therefore, if Jones's heart slows, then hurries, after the control alternatives, but hurries, then slows, after the relevant alternatives, we might conclude that Jones has guilty knowledge. Smith might show the reverse pattern and yet still betray the damning fact that he recognizes which alternatives are relevant. Note the important difference between the GKT and any lie test: If Jones's heart rate response is consistently different to the control and relevant questions on a lie test, the interpretation is hopelessly ambiguous—because, on any lie test, the difference between the control and relevant questions is always apparent to guilty and innocent subjects alike.

Another response variable that may have potential for use with the GKT is the pupillary response, a minute but reliable change in the diameter of the pupil of the eye. Modern instruments, employing a TV or video camera, can measure pupil size with exquisite precision and trace the changes on a polygraph chart. As is true with heart rate, one cannot predict how recognition of the relevant alternative will affect the direction or size of the pupillary response—but that is no handicap for the GKT. As long as recognition makes a difference, any difference, then guilty knowledge can be demonstrated. For most subjects, it is doubtful that the electrodermal response can be improved upon in respect to sensitivity. The addition of heart rate, pupil size, brain wave, or voice stress responses, however, might provide a combination of variables that could defeat any countermeasure that would be successful against the EDR alone.

Forensic Applications of the GKT

In spite of the promising indications in the scientific literature, professional polygraphers still have not taken up the Guilty Knowledge Test for use in criminal investigation. The reason given for this neglect is that the examiner seldom has available to him the necessary "keys," the items of information that only a guilty suspect would recognize and that could be turned into GKT items. The lie test, in contrast, is easy to set up. All the examiner needs to know to prepare the usual lie test question list he can learn in a few minutes' conversation with the investigator, nor does it matter if the suspect has already been extensively interrogated. And there are crimes that simply do not lend themselves to the GKT approach at all. A 1980 news item referred to a former Green Beret officer, for example, accused of killing his wife and children. He claimed that his family was attacked by a gang of "long-haired hippies" and that he somehow escaped. Since—guilty or innocent—he was present at the crime scene and is familiar with all its gruesome details, it seems unlikely that a Guilty Knowledge Test could have been constructed for use with this suspect. The Peter Reilly case would not have been suitable for the GKT either, and for similar reasons.

David Raskin (who, with his former students[2]—the "Utah Group"—collectively compose almost the only scientific support for lie detection) has for reasons of his own renamed the GKT (he calls it the "concealed information test"). Supposing as he does that the simpler technique of lie detection is incredibly accurate, Raskin apparently believes that an effective tool for detecting guilty knowledge would have little forensic utility. J. A. Podlesny, one of Raskin's former students who now does polygraph work for the FBI, produced in 1995 a Justice Department Technical Report[3] alleging to show that the GKT would not be useful in FBI criminal investigations. The fact that Podlesny refers to the GKT as a "deception detection method" suggests at the outset that he may not fully understand the method he is dealing with. To support his negative conclusion, Podlesny reports having reviewed "758 polygraph examination requests ... for the presence or absence of operable case facts which are necessary for guilty knowledge tests" (p. 2). He found that fewer than 9% of the cases now brought to the FBI polygraph unit include sufficient facts or keys to permit the examiners to create adequate GKTs. Had Podlesny been working at Scotland Yard in 1900 at the time of the introduction of the Galton–Henry system of fingerprint identification, it is likely that he would also have found very few cases in the records of the Yard that included fingerprints of suspects. Were FBI investigators to be trained to search fresh crime scenes for usable GKT items, to collect facts and photographs appropriate for this purpose, there would still be many cases where the GKT could not be used—but there would be many more than the 9% of cases where

Podlesny was able to find GKT keys even though the investigators had not systematically searched for them.

Many crimes that are the subject of well-publicized trials could have been illuminated through the use of a Guilty Knowledge Test, provided that the person responsible for constructing the test items was able to participate in the investigation from the outset. The real reason why this method is not used, I believe, is that police polygraphers are not investigators and the detectives who actually conduct the investigations do not know much about polygraphic interrogation. If the potential of the GKT is to be realized, the test constructor must be early at the crime scene, camera and notebook in hand, looking as with the eyes of the perpetrator for those keys from which good test items could be contrived. The most efficient arrangement would be for the person responsible for the test to be the same person who is responsible for the investigation itself. This combining of responsibilities should ensure that the identification of keys will get appropriate priority and also that these keys will not be vitiated later by careless questioning of suspects. This does not mean that all detectives would have to learn how to run a polygraph or read charts; the actual test administration would be done as it is now by a specialist. The detectives would learn only the psychological principles of the GKT and how to construct good items.

The GKT in the Courtroom

The Guilty Knowledge Test is being advocated here as an investigative tool and not as a method of strengthening, in the eyes of a jury, the case for or against a criminal defendant. But a defense attorney who learned that a GKT had been administered to his client, and that the score attained had been as low as that produced by the known-innocent subject employed to calibrate the test, would be sorely tempted to have that fact brought into evidence. Similarly, a prosecutor, knowing that a defendant had scored higher than would be expected in 10,000 testings of innocent suspects, might reasonably feel that this information would assist the jury in its decision making. Since there are striking differences between the GKT and lie test evidence, both in respect to probable validity and also in the intrinsic nature of the evidence, it seems worthwhile to point out some of these distinguishing features.

In the first place, while any lie test requires the defendant to reply to the basic question at issue, "Did you do it?" the GKT requires no verbal testimony at all. The suspect is never asked if he "did it." The GKT alternatives are presented in the form of questions but those questions are not answered verbally; the only vocal response consists of echoing the question. The response from which guilty knowledge is to be inferred is nonverbal and involuntary. It can be argued that requiring a defendant to submit to a Guilty Knowledge Test would be analogous to requiring him to show his face to a witness or to give a sample of his fingerprints. Requiring him to submit to a lie test, in contrast, is requiring him to testify against himself. Whether this difference makes a difference with respect to Fifth Amendment issues, my legal consultants have been uncertain about. It is, they have felt, "an interesting question."

Second, the examiner who would offer into evidence the results of a GKT would not be in the position of asking the court to accept his expert opinion as to the guilt of the accused. He would describe how the test was constructed, how it was calibrated on one or more known-innocent subjects. He would explain the score attained by the defendant and how that score was derived. Finally, he would explain the appropriate statistical interpretation of that score, the probability that such a score might be produced by a subject without guilty knowledge. The jury could be easily apprised of all the relevant facts and could as easily arrive at its own conclusions. Thus, in marked contrast to the lie detector, admitting into evidence the results of a Guilty Knowledge Test would not usurp the jury's role as finder of fact. That, it seems to me, makes a considerable difference.

Future Prospects

It should be said again that the Guilty Knowledge Test, as here described, has not been systematically studied in the context of real-life criminal investigation. Questions remain that could be addressed in further laboratory work. But the important questions—Can the GKT be useful in criminal investigation? Can it be made as accurate, especially in exonerating innocent suspects, as the theory suggests?—will only be answered by adequate field trial. Such field research cannot be done by college professors or polygraphers working alone but will require the active participation of interested police detectives willing to try something new. If this book succeeds in fomenting a healthy skepticism about the flourishing industry of lie detection, honesty testing, and the private practice of police work generally, while at the same time generating a skeptical interest in the possibilities of guilt detection for use in criminal investigation by the duly constituted authorities, then it will have accomplished its objectives.

NOTES AND REFERENCES

Chapter 1

1. States in which the results of lie detector tests are inadmissible at trial: Colorado, Hawaii, Illinois, Kentucky, Louisiana, Maine, Maryland, Michigan, Minnesota, Mississippi, Missouri, Montana, Nebraska, New Hampshire, New York, North Carolina, Oklahoma, Pennsylvania, South Dakota, Tennessee, Texas, West Virginia, and Wisconsin.
2. *Daubert v. Merrell Dow Pharmaceuticals*, 113 S.Ct. 2786 (1993); see D. L. Faigman, The evidentiary status of social science under *Daubert*: Is it "scientific," "technical," or "other" knowledge? *Psychology, Public Policy, & Law*, 1995, 1, 960–979.

Chapter 2

1. Hold the mustard, *Cincinnati Inquirer*, July 1, 1975.

Chapter 3

1. G. Steiner, *After Babel*, London: Oxford University Press, 1975.
2. Professor David Premack, University of Pennsylvania, telephone conversations, 1977.
3. In his excellent book *The Language Instinct* (New York: W. Morrow, 1994) Steven Pinker reveals the truth about the "Great Eskimo Vocabulary Hoax." It seems to have been invented by the amateur scholar of Native American languages Benjamin Lee Whorf (pp. 64–65).
4. J. A. Larson, *Lying and Its Detection*, Chicago: University of Chicago Press, 1932.
5. C. Lombroso, *L'Homme Criminel* (2nd French ed., 1895), cited in P. V. Trovillo, A

history of lie detection, *Journal of Criminal Law, Criminology and Police Science*, 1939, *30*, 848–881.

6. A. R. Luria, *The Nature of Human Conflicts,* New York: Liverright, 1932.
7. F. Galton, Psychometric experiments, *Brain*, 1879, *2*, 162.
8. W. M. Marston, *The Lie Detector Test*, New York: Richard R. Smith, 1938, p. 45.
9. Ibid., p. 81.
10. Ibid., p. 87.
11. *Look*, December 6, 1938, pp. 1–17.
12. J. G. Linehan, Lie detection pioneer profiles, *Polygraph*, 1978, *7*, 95–100.
13. Larson, *Lying*, p. 333.
14. Ibid., p. 121.
15. J. A. Larson, The lie detector polygraph: Its history and development, *Journal of the Michigan State Medical Society*, 1938, *37*, 893–897.
16. Ibid., p. 896.
17. Cited in J. H. Skolnick, Scientific theory and scientific evidence: An analysis of lie-detection, *The Yale Law Journal*, 1961, *70*, 694, 728.
18. C. D. Lee, *The Instrumental Detection of Deception: The Lie Test*, Springfield, Ill.: Charles C. Thomas, 1953.
19. J. E. Reid and F. E. Inbau, *Truth and Deception, The Polygraph ("Lie Detector") Technique*, 2nd ed., Baltimore: Williams & Wilkins, 1977.
20. J. E. Reid, A revised questioning technique in lie-detection tests, *Journal of Criminal Law and Criminology*, 1947, *37*, 542–547.
21. J. E. Reid and R. O. Arther, Behavior symptoms of lie detector subjects, *Journal of Criminal Law and Criminology*, 1953, *44*, 104–108.
22. D. T. Lykken, Psychology and the lie detector industry, *American Psychologist*, 1979, *29*, 725–739 (reprinted in M. Steinmann, Jr., *Words in Action*, New York: Harcourt Brace Jovanovich, 1979); D. T. Lykken, The right way to use a lie detector, *Psychology Today*, 1975, Vol. 8, pp. 56–60 (reprinted in Steinmann, *Words in Action*); D. T. Lykken, The lie detector industry: Just nine years more to 1984, *Modern Medicine*, 1975 (reprinted in Steinmann, *Words in Action*); D. T. Lykken, Polygraph tests in business: Unscientific, unAmerican, illegal, *Hennepin Lawyer*, 1976, *44*, 4, 28.
23. OTA Report, *Scientific Validity of Polygraph Testing: A Research Review and Evaluation*, Washington, D.C.: U.S. Congress, Office of Technology Assessment (OTA-TM-H-15, November), 1983.
24. L. Saxe, On the proposed use of polygraphs in the Department of Defense, Statement before the Committee on Armed Services of the United States Senate, presented at the request of the American Psychological Association, March 7, 1984; L. Saxe, On deceiving ourselves in detecting deceit: Espionage, science, and public policy, *Issues in Science and Technology*, 1985, *11*, 15–16; L. Saxe, D. Dougherty, and T. Cross, The validity of polygraph testing: Scientific analysis and public controversy, *American Psychologist*, 1985, *40*, 355–366.
25. J. J. Furedy, Lie detection as psychophysiological differentiation: Some fine lines. In G. H. Coles, E. Donchin, and S. W. Porges (eds.), *Psychophysiology: Systems, Processes and Applications*, New York: Guilford Press, 1986; J. J. Furedy, The North American CQT polygraph and the legal profession: A case of Canadian credulity and a cause for cultural concern, *Criminal Law Quarterly*,

1989, pp. 43–49; A. Gale (ed.), *The Polygraph Test: Lies, Truth and Science*, London: Sage, 1988.

26. B. Kleinmuntz and J. J. Szucko, On the fallibility of lie detection, *Law & Society Review*, 1982, *17*, 85–104; B. Kleinmuntz and J. J. Szucko, A field study of the fallibility of polygraphic lie detection, *Nature*, 1984, *308*, 449–450.

27. W. G. Iacono, G. A. Boisvenu, and J. A. Fleming, Effects of diazepam and methylphenidate on the electrodermal detection of guilty knowledge, *Journal of Applied Psychology*, 1984, *69*, 289–299; W. G. Iacono and C. J. Patrick, What psychologists should know about lie detection. In A. K. Hess and I. B. Weiner (eds.), *Handbook of Forensic Psychology*, New York: John Wiley, 1987; W. G. Iacono and C. J. Patrick, Polygraph techniques. In R. Rogers (ed.), *Clinical Assessment of Malingering and Deception*, New York: Guilford Press, 1988; W. G. Iacono and D. T. Lykken, The scientific status of research on polygraph tests: The case against polygraph tests. In D. L. Faigman, D. Kaye, M. J. Saks, and J. Sanders (eds.), *West Companion to Scientific Evidence*, St. Paul, Minn.: West Publishing, 1997; W. G. Iacono and D. T. Lykken, The validity of the lie detector: Two surveys of scientific opinion, *Journal of Applied Psychology*, 1997, *82*, 426–433.

28. B. Rice, The new truth machines, *Psychology Today*, June 1978, pp. 61–78.

29. R. O. Arther, *The Scientific Investigator*, Springfield, Ill.: Charles C. Thomas, 1965.

30. R. J. Ferguson and A. L. Miller, *Polygraph for the Defense*, Springfield, Ill.: Charles C. Thomas, 1974, pp. 36–37. This is by no means an isolated example. A past president of the American Polygraph Association inveighs against "the do-gooders, the bleeding hearts, and the civil libertarians ... [who] see no reason why a person who elects a life of crime should not enjoy absolute privacy to continue his career unmolested" (R. J. Weir, Jr., *Polygraph*, 1974, *3*, 125). The editor of the *APA Newsletter* (May–June 1974, p. 25) warns his readers of a State Department initiative to admit into the United States 500 political prisoners from Chile and Argentina. Referring to these refugees as "terrorists, communists, and revolutionaries," he suggests that polygraphers called upon to give preemployment tests to such persons should "take a refresher course in communism, Marxism, and terrorism." In a letter attacking me for an article opposing preemployment screening, H. M. Hanson of Scientific Security Systems in Phoenix comments: "You, not being responsible for meeting a payroll or overhead, would not and cannot understand why employers do not wish to have ... ex-convicts or 'hypes' (those on narcotics) on our payrolls, even though various bureaucratic agencies and obviously liberal colleges permit this to exist within their organizations. I have always felt that the University of Minnesota was a rather conservative and certainly highly respected university in the academic world. Obviously the likes of Humphrey and other liberals have changed the image at Minnesota" (January 8, 1975).

31. R. E. Smith, *Privacy: How to Protect What's Left of It*, Garden City, N.Y.: Anchor Press/Doubleday, 1979, p. 255.

32. E. A. Jones, Jr., "Truth" when the polygraph operator sits as arbitrator (or judge): The deception of "detection" in the "Diagnosis of truth and deception." In J. Stern and B. Dennis (eds.), *Truth, Lie Detectors, and Other Problems*, 31st Annual Proceedings, National Academy of Arbitrators, 1979, p. 133.

33. Ibid., fn. 108, p. 133.

34. J. Frank, *Courts on Trial*, Princeton, N.J.: Princeton University Press, 1949, pp. 247–335.
35. Pinocchio's new nose (editorial), *New York University Law Review*, 1973, *48*, p. 340.
36. L. M. Burkey, Privacy, property and the polygraph, *Labor Law Journal*, 1967, *18*, 79.
37. J. H. Skolnick, Scientific theory and scientific evidence: An analysis of lie detection, *Yale Law Journal*, 1961, *70*, 694–728.
38. Jones, "Truth."
39. Justice D. R. Morand, *The Royal Commission into Metropolitan Toronto Police Practices*, Toronto, Canada, 1976.
40. N. Ansley and F. Horvath, *Truth and Science*, Linthicum Heights, Md.: American Polygraph Association, 1977.

Chapter 4

1. W. M. Marston, *The Lie Detector Test*, New York: Richard R. Smith, 1938, pp. 32, 8.
2. R. W. Gerard, To prevent another world war: Truth detection, *Journal of Conflict Resolution*, 1961, *5*, 212–218.
3. J. G. Linehan, Lie detection pioneer profiles, *Polygraphy*, 1978, *7*, 95–100.
4. H. J. Rothwax, *Guilty: The Collapse of Criminal Justice*, New York: Random House, 1996.
5. U.S. Department of Justice, *Sourcebook of Criminal Justice Statistics*, Washington, D.C.: U.S. Department of Justice, 1973.
6. S. Bok, *Lying: Moral Choice in Public and Private Life*, New York: Pantheon, 1978.
7. As this book went to press, I discovered a novel by James I. Halperin (*The Truth Machine*, New York: Ballantine Books, 1996) in which the author imagines in far greater (and more interesting) detail than I have done the social, economic, and international consequences that might ensue from the development of an infallible truth verifier.
8. F. Alexander, *Psychosomatic Medicine*, New York: Norton, 1950.
9. A. F. Ax, The physiological differentiation between fear and anger in humans, *Psychosomatic Medicine*, 1953, *15*, 433–442.
10. J. Schachter, Pain, fear and anger in hypertensives and normotensives, *Psychosomatic Medicine*, 1957, *19*, 17–29.
11. J. I. Lacey and B. C. Lacey, Verification and extension of the principle of autonomic response-stereotypy, *American Journal of Psychology*, 1958, *71*, 50–73.
12. W. James, What is emotion? *Mind*, 1884, *9*, 188–205.
13. W. B. Cannon, *Bodily Changes in Pain, Hunger, Fear and Rage*, 2nd ed., New York: Appleton-Century-Crofts, 1929.
14. G. Marañon, Contribution a l'étude de l'action emotive de l'adrénaline, *Revue Français Endocrinologie*, 1924, *2*, 301–325.
15. G. W. Hohmann, Some effects of spinal cord lesions on experienced emotional feelings, *Psychophysiology*, 1966, *3*, 143–156.

16. J. E. Reid and F. E. Inbau, *Truth and Deception*, 2nd ed., Baltimore: Williams & Wilkins, 1977, pp. 61–68.

17. L. A. Geddes and D. C. Newberg, Cuff pressure oscillations in the measurement of relative blood pressure, *Psychophysiology*, 1977, *14*, 198–202.

18. I. Younger, Review of the first edition of Reid and Inbau's text, *Saturday Review*, December 31, 1966.

19. D. C. Raskin and R. D. Hare, Psychopathy and detection of deception in a prison population, *Psychophysiology*, 1978, *15*, 126–136. See also my critique: D. T. Lykken, The psychopath and the lie detector, *Psychophysiology*, 1978, *15*, 137–142.

20. R. J. Weir, Jr., In defense of the relevant-irrelevant polygraph test, *Polygraph*, 1974, *3*, 124.

Chapter 5

1. P. E. Meehl, *Clinical and Actuarial Prediction*, Minneapolis: University of Minnesota Press, 1954, p. 2.

2. J. E. Reid and F. E. Inbau, *Truth and Deception*, 2nd ed., Baltimore: Williams & Wilkins, 1977, p. 304.

3. R. O. Arther, *The Scientific Investigator*, Springfield, Ill.: Charles C. Thomas, 1965.

4. W. G. Summers, Science can get the confession, *Fordham Law Review*, 1939, *8*, 334–354.

5. Mr. Minor made this claim on the *Larry King Live* program (CNN TV) on March 21, 1997.

6. If an experienced examiner had verified, say, 10% of his cases, and if these 10% were strictly representative of all his cases, then the accuracy obtained on this sample might reasonably be generalized to all his cases. But the verified sample is never a random or representative subset of the total, and this is not the way in which these high estimates are arrived at. In Inbau and Reid's first text on lie detection, for example (*Lie Detection and Criminal Interrogation*, Baltimore: Williams & Wilkins, 1953), the error rate was computed by dividing the number of known errors by the entire number of tests administered (not just the verified ones), yielding the astonishing claim of only .07% errors (actually reported as .0007% due to a failure to convert from decimals to percentages). The more recent text does not show how the calculation was done and gives only the results.

7. See W. G. Iacono, Can we determine the accuracy of polygraph tests? In J. R. Jennings, P. A. Ackles, and M. Coles (eds.), *Advances in Psychophysiology, Vol. 4*, London: Kingsley, 1991.

8. K. E. Murray, Movement recording chairs: A necessity? *Polygraph*, 1989, *18*, 15–23. Mr. Murray was responding to a proposal that polygraph subjects should sit in specially constructed chairs designed to detect movements intended to beat the CQT. His argument was that if a police polygrapher could achieve 99.4% accuracy, as he (mistakenly) believed his data showed, then such chairs were unnecessary.

9. S. Abrams, *A Polygraph Handbook for Attorneys*, Lexington, Mass.: Lexington Books, 1977.

10. C. N. Parkinson, *Parkinson's Law and Other Studies in Administration*, Boston: Houghton Mifflin, 1957.

11. L. J. Peter, *The Peter Principle*, New York: Morrow, 1969.

12. F. S. Horvath, The effect of selected variables on interpretation of polygraph records, *Journal of Applied Psychology*, 1977, *62*, 127–136.

13. G. Barland and D. Raskin, *Validity and reliability of polygraph examinations of criminal suspects* (Report No. 76-1, Contract 75-NI-99-0001), Washington, D.C.: U.S. Department of Justice, 1976.

14. M. E. Bitterman and F. L. Marcuse, Cardiovascular responses of innocent persons to criminal investigations, *American Journal of Psychology*, 1947, *60*, 407–412.

15. C. J. Patrick and W. G. Iacono, Validity of the control question polygraph test: The problem of sampling bias, *Journal of Applied Psychology*, 1991, *76*, 229–238.

16. This was first demonstrated by Iacono, Can we determine the accuracy of polygraph tests?

Chapter 6

1. J. E. Reid and R. O. Arther, Behavior symptoms of lie detector subjects, *Journal of Criminal Law and Criminology*, 1953, *44*, 104–108.

2. J. E. Reid and F. E. Inbau, *Truth and Deception*, 2nd ed., Baltimore: Williams & Wilkins, 1977, pp. 293–295.

3. Quoted in the report of the *Royal Commission into Metropolitan Toronto Police Practices*, 1976, Hon. Mr. Justice Donald R. Morand, Commissioner, p. 247.

4. Reid and Inbau, *Truth and Deception*, p. 42, n. 49.

5. D. Van Buskirk and F. Marcuse, The nature of errors in experimental lie detection, *Journal of Experimental Psychology*, 1954, *47*, 187–190; L. A. Gustafson and M. T. Orne, Effects of heightened motivation on the detection of deception, *Journal of Applied Psychology*, 1963, *47*, 408–411; S. Kugelmass *et al.*, Experimental evaluation of galvanic skin response and blood pressure change indices during criminal interrogation, *Journal of Criminal Law, Criminology and Police Science*, 1968, *59*, 632–635.

6. Morand, *Royal Commission*, p. 244.

7. Reid and Arther, Behavior symptoms.

8. F. Horvath, Verbal and nonverbal clues to truth and deception during polygraph examinations, *Journal of Police Science and Administration*, 1973, *1*, 138–152.

9. E. A. Jones, Jr., Evidentiary concepts in labor arbitration: Some modern variations on ancient legal themes, *U.C.L.A. Law Review*, 1966, *13*, 1286.

10. Morand, *Royal Commission*, p. 245.

11. J. Kubis, *Comparison of Voice Analysis and Polygraph as Lie Detection Procedures* (Contract DAAD05-72-C-0217), U.S. Army Land Warfare Laboratory, Aberdeen Proving Ground, Maryland, 1973; G. Barland, *Detection of Deception in Criminal Suspects*, Doctoral dissertation, University of Utah, 1975.

12. P. Ekman, *Telling Lies: Clues to Deceit in the Marketplace, Politics, and Marriage*, New York: W. W. Norton, 1992, p. 285.
13. P. M. Bersh, A validation study of polygraph examiner judgements, *Journal of Applied Psychology*, 1969, 53, 399–403.
14. Morand, *Royal Commission*, p. 261.

Chapter 7

1. J. A. Larson, The lie detector polygraph: Its history and development, *Journal of the Michigan State Medical Society*, 1938, 37, 893–897. A number of studies have been reported in which a large group of suspects were tested by the R/I method in relation to the same crime. In every instance except for the cited study by Larson, the persons who scored the charts were aware that not more than one person could reasonably be guilty and therefore the scorers could have achieved very high "accuracy" just by calling everyone truthful. Thus, Bitterman and Marcuse tested 81 residents of a college dormitory where $100 had been stolen from a student's room. Finding that 7 of 81 students "failed" the R/I test the first time around, Bitterman and Marcuse retested those 7 and finally concluded that all of them were innocent (M. E. Bitterman and F. L. Marcuse, Cardiovascular responses of innocent persons to criminal investigation, *American Journal of Psychology*, 1947, 60, 407–412). The only useful evidence of lie test accuracy is obtained when the chart evaluator reads each chart independently with no outside reason for expecting either a truthful or a deceptive answer.
2. F. Horvath, The utility of control questions and the effects of two control question types in field polygraph techniques, *Journal of Police Science and Administration*, 1968, 16, 357–379.
3. S. W. Horowitz, J. C. Kircher, C. R. Honts, and D. C. Raskin, The role of comparison questions in physiological detection of deception, *Psychophysiology*, 1997, 34, 108–115.

Chapter 8

1. *Proceedings at trial, Queen v. William Wong*, Supreme Court of British Columbia, No. CC760628, Vancouver, B.C., Canada, October 1976.
2. D. Raskin, The polygraph in 1986: Scientific, professional and legal issues surrounding applications and acceptance of polygraph evidence, *Utah Law Review*, 1986, 1, 29–74.
3. D. Raskin, Polygraph techniques for the detection of deception. In D. Raskin (ed.), *Psychological Methods in Criminal Investigation and Evidence*, New York: Springer, 1989, pp. 247–296.
4. Lie test for rape victims, *Mother Jones*, July 1979, p. 10. This practice is widespread in the United States. Jan Leventer, codirector of the Women's Justice Center (WJC) in Detroit, determined in 1978 that rape victims were being

subjected to polygraph tests in at least 17 states. The WJC brought suit to stop the practice in Detroit and, in January of 1979, the Detroit Police Department agreed to these demands. In Maryland, one of the "relevant" questions routinely asked during lie tests of rape victims was "Did you have an orgasm?"

  5. C. Backster, Anticlimax dampening concept, *Polygraph*, 1974, *3*, 28–50.
  6. D. Raskin, Scientific assessment of the accuracy of detection of deception: A reply to Lykken, *Psychophysiology*, 1978, *15*, 143–147.
  7. *Arizona v. Pete*, Superior Court of Arizona, No. CR-77905, October 1974.
  8. F. Horvath and J. Reid, The reliability of polygraph examiner diagnosis of truth and deception, *Journal of Criminal Law, Criminology and Police Science*, 1971, *62*, 276–281.
  9. F. Hunter and P. Ash, The accuracy and consistency of polygraph examiner's diagnosis, *Journal of Police Science and Administration*, 1973, *1*, 370–375.
 10. D. Wicklander and F. Hunter, The influence of auxiliary sources of information in polygraph diagnosis, *Journal of Police Science and Administration*, 1975, *3*, 405, 409.
 11. S. Slowick and J. Buckley, Relative accuracy of polygraph examiner diagnosis of respiration, blood pressure, and GSR recordings, *Journal of Police Science and Administration*, 1975, *3*, 305–309.
 12. D. C. Raskin, J. C. Kircher, C. R. Honts, and S. W. Horowitz, *A Study of the Validity of Polygraph Examinations in Criminal Investigation*, final report to the National Institute of Justice (Grant No. 85-IJ-CX-0040), Department of Psychology, University of Utah, May 1988.
 13. J. Kircher, S. Horowitz, and D. Raskin, Meta-analysis of mock crime studies of the Control Question polygraph technique, *Law and Human Behavior*, 1988, *12*, 79–90.
 14. Ibid.
 15. Ibid.
 16. C. Patrick and W. G. Iacono, Psychopathy, threat, and polygraph test accuracy, *Journal of Applied Psychology*, 1989, *74*, 347–355 (specifically pp. 348–349).
 17. Ibid., p. 350, Table 1.
 18. R. F. Forman and C. McCauley, Validity of the positive control test using the field practice model, *Journal of Applied Psychology*, 1986, *71*, 691–698.
 19. Ibid., p. 693.
 20. F. Horvath, The effect of selected variables on interpretation of polygraph records, *Journal of Applied Psychology*, 1977, *62*, 127–136; B. Kleinmuntz and J. Szucko, A field study of the fallibility of polygraphic lie detection, *Nature*, 1984, *308*, 449–450; C. Patrick and W. G. Iacono, Validity of the control question polygraph test: The problem of sampling bias, *Journal of Applied Psychology*, 1991, *76*, 229–238; C. Honts, Criterion development and validity of the CQT in field application, *Journal of General Psychology*, 1996, *123*, 309–324.
 21. G. H. Barland and D. C. Raskin, *Validity and Reliability of Polygraph Examinations of Criminal Suspects* (Report 76-1, Contract 75 NI-99-000), Washington, D.C.: U.S. Department of Justice, 1976.
 22. Justice D. R. Morand, in *Royal Commission into Metropolitan Toronto Police Practices*, 1976, p. 22.

Chapter 9

1. D. Raskin, Polygraph techniques for the detection of deception. Chapter 8 in D. Raskin (ed.), *Psychological Methods in Criminal Investigation and Evidence*, New York: Springer, 1989, p. 271.
2. Ibid., p. 254.
3. S. W. Horowitz, J. C. Kircher, C. R. Honts, and D. C. Raskin, The role of comparison questions in physiological detection of deception, *Psychophysiology*, 1997, *34*, 108–115.
4. J. Kircher, S. Horowitz, and D. Raskin, Meta-analysis of mock crime studies of the Control Question polygraph technique, *Law and Human Behavior*, 1988, *12*, 79–90.
5. This actually happened to a Washington State grandfather who went to trial after "failing" the lie detector. According to his attorney, with whom I corresponded, his client was saved by the testimony of his physician that, as a consequence of surgery, the old man had been incapable of having an erection for ten years, and this proved that the granddaughter's graphic accusations were false.
6. One study of the DLT by the Raskin group has appeared, but not in a peer-reviewed scientific journal: C. R. Honts and D. C. Raskin, A field study of the validity of the Directed Lie control question, *Journal of Police Science and Administration*, 1988, *161*, 56–61.
7. R. F. Forman and C. McCauley, Validity of the Positive Control Test using the field practice model, *Journal of Applied Psychology*, 1986, *71*, 691, 698.
8. Ibid.
9. J. E. Reid, A revised questioning technique in lie-detection tests, *Journal of Criminal Law and Criminology*, 1947, *37*, 542–547.
10. The mean difference between reactions to the relevant and the known-truth questions was zero for the innocent subjects. One reaction was larger for 9 of the 20 subjects and the other was larger for 11. Whether all of these subjects would have been scored as truthful is unclear from the report: J. Podlesny and D. Raskin, Effectiveness of techniques and physiological measures in the detection of deception, *Psychophysiology*, 1978, *15*, 344–359.

Chapter 10

1. From a letter to the editor of *JAMA* (*Journal of the American Medical Association*), January 9, 1987, p. 190.
2. This review originally appeared in *Contemporary Psychology*, 1985, *30*, 880–881.
3. C. Gugas, *The Silent Witness*, Englewood Cliffs, N.J.: Prentice-Hall, 1979.
4. Permanent Select Committee on Intelligence, U.S. House of Representatives, News Release, September 24, 1979.

Chapter 11

1. Use of polygraphs as "lie detectors" by federal agencies, *Hearings before a Subcommittee of the Committee on Government Operations*, U.S. House of Representatives, 88th Congress, 1964.
2. O. Lippold, Physiological tremor, *Scientific American*, 1971, 224, 65–73.
3. G. Inbar and G. Eden, Psychological Stress Evaluators: EMG correlation with voice tremor, *Biological Cybernetics*, 1976, 24, 165–167.
4. M. Brenner, *Stagefright and Steven's Law*, Presentation to the Eastern Psychological Association, April 1974.
5. G. Smith, Voice analysis for the measurement of anxiety, *British Journal of Medical Psychology*, 1977, 50, 367–373.
6. M. Brenner, H. Branscomb, and G. Schwartz, Psychological Stress Evaluator—Two tests of a vocal measure, *Psychophysiology*, 1979, 16, 351–357.
7. B. Lynch and D. A. Henry, Validity study of the Psychological Stress Evaluator, *Canadian Journal of Behavioral Science*, 1979, 11, 89–94.
8. I. Nachshon, *The Psychological Stress Evaluator: Validity Study*, final report (Grant #953-0265-001), Israeli Police, Department of Criminology, Bar Ilan University, Ramat Gan, Israel.
9. Polygraph control and Civil Liberties Protection Act, *Hearings before the Subcommittee on the Constitution of the Committee on the Judiciary*, United States Senate, September 19, 1978.
10. Ibid.
11. J. Kubis, Comparison of voice analysis and polygraph as lie detection procedures, *Polygraph*, 1974, 3, 1–47.
12. Nachshon, *The Psychological Stress Evaluator*.
13. F. Horvath, Effect of different motivation instructions on detection of deception with the PSE and the GSR, *Journal of Applied Psychology*, 1979, 64, 323–330.
14. D. T. Lykken, The validity of the guilty knowledge technique: The effects of faking, *Journal of Applied Psychology*, 1960, 44, 258–262.
15. Brenner *et al.*, Psychological Stress Evaluator.
16. G. Barland, Use of voice changes in the detection of deception, *Polygraph*, 1978, 7, 129–140.
17. Y. Tobin, A validation of voice analysis techniques for the detection of psychological stress, Unpublished paper, cited in Nachshon, *The Psychological Stress Evaluator*.
18. Nachshon, *The Psychological Stress Evaluator*, p. 49.
19. This description of the "Truth Phone's" capabilities is quoted from www.spy-zone.com/CCSDT1.html#CCSDT6.
20. This statement is quoted from a letter from Dr. Humble dated May 14, 1997.
21. DoDPI Voice Stress Analysis position statement, dated September 11, 1996 and signed by Michael H. Capps, Director.
22. Reported by e-mail from Linda J. Quinones, Deputy Sheriff and polygraph examiner in the Los Angeles County Sheriff's Department and president of the California State Association of Polygraphists, dated May 15, 1997.

Chapter 12

1. *Daubert v. Merrell Dow Pharmaceuticals*. See D. L. Faigman, The evidentiary status of social science under *Daubert*: Is it "scientific," "technical," or "other" knowledge? *Psychology, Public Policy, & Law*, 1995, 1, 960–979.
2. D. L. Faigman, D. Kaye, M. J. Saks, and J. Saunders (eds.), *Modern Scientific Evidence: The Law and Science of Expert Testimony*, St. Paul, Minn.: West Publishing, 1997.
3. S. L. Amato and C. R. Honts, What do psychophysiologists think about polygraph tests? A survey of the membership of SPR, *Psychophysiology*, 1993, 31, S22 (Abstract).
4. Ibid.
5. S. S. Diamond, Reference guide to survey research. In Federal Judicial Center *Reference Manual on Scientific Evidence*, Washington, D.C.: Clark Boardman Callaghan Publishing, 1995, pp. 223–271.
6. Much of this chapter is adapted, with permission, from W. G. Iacono and D. T. Lykken, The Validity of the lie detector: Two surveys of scientific opinion, *Journal of Applied Psychology*, 1997, 82, 426–433.
7. D. Raskin, The polygraph in 1986: Scientific, professional and legal issues surrounding applications and acceptance of polygraph evidence. *Utah Law Review*, 1986, 1, 29–74.
8. D. Raskin, Polygraph techniques in the detection of deception. In D. Raskin (ed.), *Psychological Methods in Criminal Investigation and Evidence*, New York: Springer, 1989, pp. 247–296.
9. C. R. Honts, R. L. Hodes, and D. C. Raskin, Effects of physical countermeasures on the physiological detection of deception, *Journal of Applied Psychology*, 1985, 70, 177–187; C. R. Honts, D. C. Raskin, and J. C. Kircher, Mental and physical countermeasures reduce the accuracy of polygraph tests, *Journal of Applied Psychology*, 1994, 79, 252–259.
10. Raskin, Polygraph techniques.
11. We did not send the SPR questionnaires to Raskin, Honts, Kircher, or ourselves, nor did we include ourselves in the mailing to Division One Fellows.
12. Diamond, Reference guide to survey research.

Chapter 13

1. J. A. Matte, *Forensic Psychophysiology Using the Polygraph*, Williamsville, N.Y.: JAM Publications (A division of Matte Polygraph Service, Inc.), 1997.

Chapter 14

1. J. Kirk Barefoot, Testimony on behalf of the American Polygraph Association, *Hearings before the Subcommittee on the Constitution of the Committee on the Judiciary*, United States Senate, 95th Congress, 1977–78, p. 60.

2. Except for the Coker family and Roach, the polygrapher, the names of the participants in this affair have been disguised.
3. Employee Polygraph Protection Act, Section 7-(d)-(1).

Chapter 15

1. This example is quoted from an article by Stephen Budiansky, the Washington editor of the British science journal *Nature*, that appeared in *The Atlantic Monthly* for October 1984.
2. Pillsbury described his unhappy experience with the FBI's polygraphic "leak detector" in an article in the *Washington Post* (November 10, 1991) as a warning to other officials about to be "fluttered" in the search for the leaker of Anita Hill's FBI report, on orders from then-President George Bush.
3. Deposition of James K. Murphy submitted March 3, 1995, to the U.S. district court for the District of New Mexico, in re *United states v. William Edward Galbreth* (Criminal No. 94-197, 26 U.S.C./7201).

Chapter 16

1. S. O. Lilienfeld, B. P. Andrews, and E. F. Stone-Romero, The relations between a self-report honesty test and personality measures in prison and college samples, *Journal of Research in Personality*, 1994, *28*, 154–169.
2. References American Management Association, *Crimes against Business: Background, Findings and Recommendations*, New York: AMA, 1977; J. P. Clark and R. C. Hollinger, *Theft by Employees in Work Organizations: Executive Summary*, Washington, D.C.: National Institute of Justice.
3. U.S. Congress, Office of Technology Assessment, *The Use of Integrity Tests for Pre-Employment Screening* (OTA SET-443), Washington, D.C.: U.S. Government Printing Office, 1990.
4. L. R. Goldberg, J. Grenier, R. Guion, L. Sechrest, and H. Wing, *Questionnaires Used in the Prediction of Trustworthiness in Pre-Employment Selection Decisions*, Washington, D.C.: American Psychological Association, 1991.
5. Violence research is due for attention, *Journal of NIH Research*, 1992, *4*, 38.
6. M. R. Gottfredson and T. Hirschi, *A General Theory of Crime*, Stanford, Calif.: Stanford University Press, 1990; J. Q. Wilson and R. L. Herrnstein, *Crime and Human Nature*, New York: Simon & Schuster, 1985.
7. Bureau of the Census, *Current Population Reports: Population Characteristics; Series P-20, No. 447, Household and Family Characteristics*, Washington, D.C.: U.S. Government Printing Office, 1990.
8. V. R. Fuchs and D. M. Reklis, America's children: Economic perspectives and policy options, *Science*, 1992, *255*, 41–46.
9. A thorough (and favorable) treatment of integrity testing can be found in D. S. Ones and C. Viswesvaren (in press). In D. W. Griffin, A. O'Leary-Kelley, and J.

M. Collins (eds.), *Dysfunctional Behavior in Organizations, Vol. 2: Nonviolent Behavior in Organizations*, Greenwich, Conn.: JAI Press; P. R. Sackett, Honesty testing for personnel selection, *Personnel Administrator*, September 1985, pp. 67–76.

10. G. E. Paajanen, *The Prediction of Counterproductive Behavior by Individual and Organizational Variables*, Unpublished doctoral dissertation, Department of Psychology, University of Minnesota, Minneapolis, 1988, p. 64.
11. Sackett, Honesty testing.
12. Goldberg *et al.*, *Questionnaires*, p. 25.
13. Goldberg *et al.*, *Questionnaires*, pp. 25–26.
14. Goldberg *et al.*, *Questionnaires*, p. 26, emphasis added.
15. G. M. Alliger, S. O. Lilienfeld, and K. E. Mitchell, The susceptibility of overt and covert integrity tests to coaching and faking, *Psychological Science*, 1996, 7, 32–39.
16. H. O. Gough, The assessment of wayward impulse by means of the Personnel Reaction Blank, *Personnel Psychology*, 1972, 24, 669–677.
17. G. M. Alliger and S. A. Dwight (in press), A meta-analytic investigation of the effect of response distortion on integrity test scores, *Psychological Assessment*.
18. Goldberg *et al.*, *Questionnaires*, p. 7; P. R. Sackett, Integrity testing for personnel selection, *Current Directions in Psychological Science*, 1994, 3, 73–76.

Chapter 17

1. C. Romig, Improving police selection with the polygraph technique, *Polygraph*, 1972, 1, 207–220.
2. D. Fox, Screening police applicants, *Polygraph*, 1972, 1, 80–82.
3. B. Skinner, *Particulars of My Life*, New York: Knopf, 1976.
4. L. Snyder, Criminal interrogation with the lie detector: Eight years experience with the Michigan State Police, *Polygraph*, 1978, 7, 79–88.
5. Selwyn Raab, Rejected confession raises questions on lie-detector use, *New York Times*, November 1, 1981.
6. E. Borchard, *Convicting the Innocent*, Garden City, N.Y.: Garden City Publishing Co., 1932.
7. R. A. Leo, False confessions and miscarriages of justice *today*, Paper presented at a conference organized by The Justice Committee, Salem, Mass., January 1997.
8. G. H. Gudjonsson, *The Psychology of Interrogations, Confessions, and Testimony*, New York: John Wiley & Sons, 1992; C. R. Huff, A. Rattner, and E. Sagarin, *Convicted but Innocent: Wrongful Conviction and Public Policy*, Thousand Oaks, Calif.: Sage Publications, 1996; M. Radelet, H. Bedau, and C. Putnam, *In Spite of Innocence: Erroneous Convictions in Capital Cases*, Boston: Northeastern University Press, 1992; L. Wrightsman and S. Kassin, *Confessions in the Courtroom*, Newbury Park, Calif.: Sage Publications, 1993; M. Yant, *Presumed Guilty: When Innocent People Are Wrongly Convicted*, Buffalo: Prometheus Books, 1991.

9. J. Barthel, *A Death in Canaan*, New York: E.P. Dutton, 1976.
10. These remarks are quoted from the Presidential Address of one Major Sylvester, at the 1910 Meetings of the International Association of Chiefs of Police, cited in J. Larson, *Lying and Its Detection*, Chicago: University of Chicago Press, 1932, pp. 97–98.
11. Since World War II, military and criminal interrogation techniques have become quite sophisticated. Fred Inbau and John Reid, in addition to their contributions to polygraphy, wrote one of the several textbooks in this field, *Criminal Interrogation and Confessions*, Baltimore: Williams & Wilkins, 1967. If jurors were to read this text, they might better understand why even innocent suspects sometimes confess to things they did not do.
12. Patt Derian, Embassy scandal was fiction, *Minneapolis Star-Tribune*, January 31, 1988. Derian was assistant secretary of state for human rights during the Carter administration.

Chapter 18

1. In N. Ansley, *Legal Admissibility of the Polygraph*, Springfield, Ill.: Charles C. Thomas, 1975.
2. *Frye v. United States*, 293 F.1013 (1924).
3. States that admit stipulated polygraph tests into evidence include Alabama, Arkansas, California, Delaware, Florida, Georgia, Indiana, Iowa, Kansas, Nevada, New Jersey, North Dakota, Ohio, Oregon, Utah, Washington, and Wyoming.
4. L. Burkey, Privacy, property, and the polygraph, *Labor Law Journal*, 1967, 18, 79.
5. D. L. Faigman, D. Kaye, M. J. Saks, and J. Sanders, *The West Companion to Scientific Evidence*, St. Paul, Minn.: West Publishing, 1997, p. 558.
6. *Tafoa v. Baca*, 402 P.2d 1001 (1985).
7. *Daubert v. Merrell Dow Pharmaceuticals*, 113 S.Ct. 2786 (1993).
8. Ibid., note 6, page 6-B.
9. Ibid., note 6, pp. 8, 9.
10. Ibid., note 6, p. 9.
11. E. Jones, "Truth" when the polygraph operator sits as arbitrator (or judge). In J. Stern and B. Dennis (eds.), *Truth, Lie Detectors, and Other Problems*, 31st Annual Proceedings, National Academy of Arbitrators, 1979, p. 151.
12. These studies are cited in the notes to Chapter 8.
13. For these calculations I have attributed to the polygraph screening test the same validity estimated for the CQT since there is no evidence at all as to the validity of the Relevant Control Test or the Positive Control Test, the only lie tests that can be used for screening purposes.
14. M. Orne, Implications of laboratory research for the detection of deception, *Polygraph*, 1975, 2, 169–199.
15. The lure of witness fees can be compelling. In 1996, polygrapher David Raskin was charging $350 an hour plus expenses.

16. R. Rosenthal, *Experimenter Effects in Behavioral Research*, New York: Appleton-Century-Crofts, 1966.
17. Jones, "Truth", p. 78.
18. A. Cimerman, "They'll let me go tomorrow": The Fay case, *Criminal Defense*, 1981, *8*(3), 7–10.
19. *People v. Kenny*, 167 Misc.51, 3 N.Y.S.2d 348 (Queens County Ct. 1938).
20. R. Koffler, The lie detector: A critical appraisal of the technique as a potential undermining factor in the judicial process, *New York Law Forum*, 1957, *3*, 123, 138–146.
21. A. Cavoukian, The effect of polygraph evidence on people's judgments of guilt, Paper presented at the meetings of the Canadian Psychological Association, June 13–15, 1979.
22. S. Carlson, M. Pasano, and J. Jannuzzo, The effect of lie detector evidence on jury deliberations: An empirical study, *Journal of Police Science and Administration*, 1977, *5*, 148–154.
23. F. Barnett, How does a jury view polygraph examination results? *Polygraph*, 1973, *2*, 277.
24. Jones, "Truth," p. 119.
25. H. Cleckley, *The Mask of Sanity*, 5th ed., St. Louis: C.V. Mosby, 1976, p. 341.
26. R. Hare, *Psychopathy: Theory and Research*, New York: John Wiley, 1970; see also D. Lykken, A study of anxiety in the sociopathic personality, *Journal of Abnormal and Social Psychology*, 1957, *55*, 6–10.
27. D. Raskin and R. Hare, Psychopathy and the detection of deception in a prison population, *Psychophysiology*, 1978, *15*, 126–136. For a critique of the above see D. Lykken, The psychopath and the lie detector, *Psychophysiology*, 1978, *15*, 137–142.
28. D. Barland and D. Raskin, Detection of deception. In W. Prokasy and D. Raskin (eds.), *Electrodermal Activity in Psychological Research*, New York: Academic Press, 1973, p. 445.
29. Gilbert & Sullivan's *The Pirates of Penzance*.
30. Letter from Professor Hare, May 25, 1978.

Chapter 19

1. From his Introduction to *A Death in Canaan*, by J. Barthel (New York: E. P. Dutton, 1976).
2. D. Lykken, I. Macindoe, and A. Tellegen, Perception: Autonomic response to shock as a function of predictability in time and locus, *Psychophysiology*, 1972, *9*, 318–333; D. Lykken and A. Tellegen, On the validity of the preception [perception?] hypothesis, *Psychophysiology*, 1974, *11*, 125–132.
3. S. Kugelmass and I. Lieblich, Relation between ethnic origin and GSR reactivity in psychophysiological detection, *Journal of Applied Psychology*, 1968, *52*, 158–162.
4. G. H. Gudjonsson, How to defeat polygraph tests. In A. Gale (ed.), *The Polygraph Test: Lies, Truth, and Science*, London: Sage, 1988.

5. J. Reid and F. Inbau, *Truth and Deception*, 2nd ed., Baltimore: Williams & Wilkins, 1977, p. 207. The examples of "respiration deception responses," on pp. 61–66 of this text, provide useful hints for persons hoping to beat the lie test. Self-induced during the control questions, such reactions will lead most examiners astray.
6. Letter to the editor of *Student Lawyer*, October 1979, responding to an article critical of polygraphy, "Bloodless Executioners," by John Jenkins, in the May 1979 issue of that journal.
7. J. Reid, Simulated blood pressure responses in lie-detector tests and a method for their detection, *American Journal of Police Science*, 1945, 36, 202–203. It should be admitted that the clinical lie test will be much harder to beat than the polygraphic lie test since examiners like Reid put greater weight on their clinical impressions, suspicions, and intuitions than on the polygraph records themselves.
8. C. R. Honts, R. L. Hodes, and D. C. Raskin, Effects of physical countermeasures on the physiological detection of deception, *Journal of Applied Psychology*, 1985, 70, 177–187; C. R. Honts, D. C. Raskin, and J. C. Kircher, Mental and physical countermeasures reduce the accuracy of polygraph tests, *Journal of Applied Psychology*, 1994, 79, 252–259.
9. Honts *et al.*, Mental and physical countermeasures, pp. 253–254.
10. Ibid., p. 252.

Chapter 20

1. H. Münsterberg, *On the Witness Stand*, New York: McClure, 1908.
2. D. Lykken, The GSR in the detection of guilt, *Journal of Applied Psychology*, 1959, 43, 385–388.
3. D. Lykken, The validity of the guilty knowledge technique: The effects of faking, *Journal of Applied Psychology*, 1960, 44, 258–262.
4. P. O. Davidson, Validity of the guilty knowledge technique: The effects of motivation, *Journal of Applied Psychology*, 1969, 53, 399–403.
5. G. Ben Shakhar *et al.*, Guilty knowledge technique: Application of signal detection measures, *Journal of Applied Psychology*, 1970, 54, 409–413 (20 items, 77% hits); W. M. Waid *et al.*, Effects of attention, as indexed by subsequent memory, on electrodermal detection of information, *Journal of Applied Psychology*, 1978, 63, 728–733 (5 items, 71%, 76%, & 77% hits).
6. Waid, Effects of attention.
7. M. Giesen and M. Rollinson, Guilty knowledge versus innocent associations: Effects of trait anxiety and stimulus context on skin conductance, *Journal of Research in Personality*, 1980, 14, 1–11.
8. I. Lieblich *et al.*, Efficiency of GSR detection of information with repeated presentation of series of stimuli in two motivational states, *Journal of Applied Psychology*, 1974, 59, 113–115.

9. J. Podlesny and D. Raskin, Effectiveness of techniques and physiological measures in the detection of deception, *Psychophysiology*, 1978, *15*, 344–359.
10. E. Elaad, A. Ginton, and N. Jungman, Detection measures in real-life guilty knowledge tests, *Journal of Applied Psychology*, 1992, *77*, 757–767; E. Elaad, Detection of guilty knowledge in real life criminal investigations, *Journal of Applied Psychology*, 1990, *75*, 521–529.
11. C. R. Honts, M. Winbush, and M. K. Devitt, *Physical and Mental Countermeasures Can Be Used to Defeat Guilty Knowledge Tests*. Poster presented at the annual meeting of the Society for Psychophysiological Research (Abstract published in *Psychophysiology*, 1994, *31*, S57). See also D. T. Lykken, The validity of the guilty knowledge technique.
12. L. A. Farwell and E. Donchin, The truth will out: Interrogative polygraphy ("lie detection") with event related potentials, *Psychophysiology*, 1991, *28*, 531–547; J. J. Allen, W. G. Iacono, and K. D. Danielson, The identification of concealed memories using the event-related potential and implicit behavioral measures: A methodology for prediction in the face of individual differences, *Psychophysiology*, 1992, *29*, 504–522.

Chapter 21

1. D. Lykken, The validity of the guilty knowledge technique: The effects of faking, *Journal of Applied Psychology*, 1960, *44*, 258–262.
2. David Raskin's former students at the University of Utah include G. H. Barland, C. R. Honts, R. L. Hodes, S. W. Horowitz, J. C. Kircher, and J. A. Podlesny.
3. J. A. Podlesny, A lack of operable case facts restricts applicability of the Guilty Knowledge Deception Detection Method in FBI criminal investigations, FBI Technical Report, Quantico, Va., 1995.

INDEX

Abrams, Stanley, 73–74
Accuracy, *see also* Reliability; Validity
   accuracy claimed by polygraphers, 64,
      69–73
   the Murray accuracy study, 71–73
Alexander, Franz, 60
Amato, S. L., 177–178
American Polygraph Association, 66, 111
American Psychological Association, 182,
      185, 226
Ames, Aldrich, 3, 68
Ansley, Norman, 50
Anticlimax dampening concept, 123
Anxiety IQ, 268
Aptitude tests, 74–75
Army Criminal Investigation Division, 104
Arther, Richard O., 30, 44–45, 69
Astrology, "proof" of validity, 129
Autonomic response to emotions, 60–62
Ax, Albert, 60

Backster, Cleve, 30, 32–34, 123–124
   Anticlimax dampening concept, 123
   Zone of Comparison method, 33
Bailey, F. Lee, 253, 258, 276
Barland, Gordon, 51, 82, 133, 170
Barthel, Joan, 240
Base rates, 256–259
Beating the lie detector: *see* Countermea-
      sures
Behavioral symptoms, 31–32, 64, 95–98
in clinical lie tests, 96
in the Control Question Test, 270

Bell, Allan, 41, 165
Bell Telephone Company, Haskins Labora-
      tory, 165
Bennett, Richard H. Jr., 171
Ben-Shakhar, Gershon, 51
Benussi, Vittorio, 64
Bersh, Paul, 104–106
Bitterman, N., 83
Blood pressure
   pletheysmograph measurement of, 65
   in polygraph "cardio" tracings, 12
Bok, Sissela, 59
Borchard, Eugene, 239–240
Bracy, Cpl. A., 245
Branscomb, Henry, 166
Brenner, Mark, 166–167
Buckley, Julian, 129
Bull Cook Book, 152–153
Burden of proof, 102, 136, 142, 159, 213, 279
Burkey, Lee M, 48, 250

Cannon, Walter, 61
Case examples
   brother accused of raping sister, 125
   CBS's stolen camera test, 197–198
   Coker vs. Piggly Wiggly, 200–205
   DeLorean, John, 69
   Demjanjuk, John, 38–40
   eating of polygraph charts, 17
   father accused of incest, 125, 251–252
   Fay murder case, 264–267
   Frye murder case, 250
   Galloway murder case, 143

327

Case examples (*cont.*)
  John K., the deputy sheriff and the bank
    robber, 206–207
  Johnson, M., false confession, 239
  Ken Chiu murder case, 116, 125
  Larson's shop-lifting case, 27
  Linda K. vs. Kresge Company, 206
  Major C., 214–216
  man who thought he'd "blacked out,"
    99
  Mary St.Clair murder case, 93–95
  Mendoza murder case, 262–264
  Mother accused of sexual abuse, 252–253
  Peter Reilly murder case, 99
  sabotage in a bakery, 199
  Sam K. and "rape" of Mary V., 125–128
  Simpson, O.J., murder case, 179, 296–299
  Sister Terressa and honesty test, 223, 225
  theatre manager, 200
  Walter K. and stolen camera, 208–210
  Wayne K. and bank theft. 205–206
Central Intelligence Agency, 103, 214
Chicago Police Department, 29
Chimpanzees, deception by, 22
Cimerman, A., 264
Cleckley, Hervey, 267
Clinical lie test, 93–114
  assumptions of, 98–104
  behavior symptoms in, 101–103
  Bersh's study of, 104–105
  posttest interrogation, 235–245
  validity of, 104–106
Computer scoring of lie tests, 79
Computer Voice Stress Analyzer (CVSA),
    41, 171–172
Confession, 235–248
  in clinical lie tests, 97
  how confessions mislead polygraphers,
    70–73
  and the courts, 245–247
  of crimes not committed, 247
  in criminal investigation, 271
  as a criterion of ground truth, 70–74,
    85–87
  fourth degree in obtaining, 235–248
  in polygraph testing, 70–74
  torture in obtaining, 23–24, 28–29,
    242–243
Consent forms, 13
Control questions, 31, 116–117

Control Question Test (CQT), 116–135, 190
  assumptions of, 119–124
  base rates in, 256–259
  examples, 116, 163
  a genuine known-lie control, 117–119
  known-lie control question, 116–117,
    120–121
  for rape victims, 120
  scientific opinion of, 183–187
  validity of, 189–191
    laboratory evidence, 84, 132–133
    real life evidence, 85–87
      good studies, 133–134
      poor studies, 128–132, 135
Control questions
  in the CQT, 116–117
  in the DLT, 138
  in a genuine control question test,
    117–119
  in the GKT, 291
  in the "positive control" test, 140–141
  in the "truth control" test, 143–145
Countermeasures
  against the GKT, 292–293, 303
  against honesty tests, 230–232
  against lie detector tests, 180–181,
    273–277
  scientific opinion about, 183–187
Courts, 249–280
  confessions in, 243–245
  the *Daubert* decision and, 253–254
  GKT as evidence in, 306–307
  jury reactions to polygraph evidence,
    265
  polygraph evidence in, 250–262
  polygraphers as expert witnesses,
    260–262
Covert lie detection, 163–164
Criminal investigation, 2
  confession in, 18, 27
  CQT in, 31, 40
  GKT in, 40, 304
  searching peak of tension test in, 148
  "voice stress analyzers" in, 171
Criterion measures, 79–80, 85–87

Dahm, A.E., 167–168
Damaging admissions, 173, 236–237
Daubert v. Merrell Dow Pharmaceuticals,
    176–177, 253–254

Dean, John, 41
Deception, confessions obtained by, 244
Deception, lie tests based on, 191–193
Defense Department Polygraph Institute
    (DoDPI), 172, 295
Dektor Counterintelligence and Security,
    Inc., 165–166, 170–171
Delorean, John, 35, 69
Demjanjuk, John, 38–40
Directed Lie Test, 137–140, 181, 190
    assumptions of, 138
    scientific opinion of, 183
    validity of, 138–139
Downes, Sgt. V., 245

Ekman, P., 103
Eden, G., 166
Elaad, E., 291
Electrodermal response (EDR), 284
Embassy: see Marine Guard scandal
Emotion
    bodily changes in, 60–62
    ethnic differences in, 273
    James's theory of, 61
Employment Inventory, the, 229–230
Employee Polygraph Protection Act, 51,
    151–152, 211, 213, 225
Employee screening, 34–35, 37, 151–161
    case against, 161, 215–222, 233–234
    case for, 153–156, 214
    cost/benefit analysis of, 226, 229–229
    See also Employee Polygraph Protection
        Act , Honesty tests, Polygraph
        screening test
Employee theft, 198–212, 226
    Coker case, 200–208
    Linda K., bookkeeper, 206
Environmental Impact Statement, 226
    extent of, 226
    good management approach to, 232–233
    police investigation of, 8, 13, 211–212
    polygraph investigation of, 197–212
    precautions against, 233
    rights of employees, 211; see also Em-
        ployee Polygraph Protection Act
    Walter K. and stolen camera, 208–210
    Wayne K., bank employee, 205–206
Error, false-positive and false-negative,
    153, 222
Exclusionary rule, 243–244

Experimenter expectancy effect, 259–260
Expert opinion
    limitations of, 68–70, 187–188
    surveys of, 175–186
Expert witnesses, 256, 260–262

Faigman, D.L., 251
Fay, Floyd, 264–267, 275–276
FBI Screening, 103, 218–222
    Elizabeth H., 218–221
    former DEA agent, 221–222
Ferguson, R. J., 45, 151–156
Fidgetometer (wiggle seat), 164
"Friendly" polygraph tests, 184, 259–262
Fourth degree, 235–247
Frank, Jerome, 47–48
Frye test, the, 250, 254
Furedy, John, 37

Galton, Sir Francis, 24
Geddes, L. A., 65
Geison, M., 289
Gelb, Leo, 35
General honesty ("control") questions,
    10–11
General Series Test, 111
Global scoring, 91, 218
Ground truth, 85–87
Gugas, C., 109, 151, 153
Guilt complex question, 31, 143–147
Guilty feelings and lying, 5
Guilty Knowledge Test (GKT), 38–41,
    181–190, 283–308
    accuracy of, 290
    countermeasures against, 292–293, 303
    as courtroom evidence, 306–307
    examples of. 38–40, 285–286, 296–301
    experimental studies of, 283–288,
        290–291
    measures for use in, 292–293, 303–304
    scientific opinion of, 183
    scoring of, 301–302
Gullibility, synonyms for, 23

Hagoth Corporation, 41, 171
Hare, Robert D., 65, 270–271
Hauptmann, Bruno, 41
Hearst, Patty. 35–36
Heisse, John, W., Jr., 168
Henry, D., 167

Herter, G.L., 152–153
Hindu method of lie detection, 24
Hohmann, G., 61
Homicide, 226
Honesty, prediction of, 223–230
Honesty tests, 43–44, 223–234
    assumptions of, 224
    countermeasures against, 230–232
    dangers of, 229–230
    validity of, 227–230
Honts, C.R., 51, 130–135, 177, 180
Horowitz, S., 51, 130–132
Horvath, Frank, 51, 79, 133–134, 170
Humble, Charles, 171–172
Hunter, F., 129

Iacono, William G., 37–38, 51, 86, 134,
        177–179, 295
Inbar, G., 166
Inbau, Fred, 25, 29, 50
Independent chart scoring, 85
Informed consent, right of, 209
Infrared (IR) detectors, 164
Institute for Defense Analysis, 54
Integrity tests: see Honesty tests
Interrogation, methods of, 18, 244
Ivan the Terrible, 38–40

James, William, 61
Job applications: see Employee screening
Jones, Edgar A., Jr., 46, 48, 102, 260, 267

Keeler, Leonarde, 29–30
Kennedy, John F., assassination of, 41
Kircher, J., 51, 130–132, 173
Klinemuntz, Benjamin, 37, 133–134
Known-lie control questions, 118, 120–121
Known-truth control questions, 18–19, 31
Kradz, Lieutenant, 168
Kubis, Joseph, 169
Kugelmass, S., 51

Lamb, R, 245
Larson, John A., 23, 27–29
Law Enforcement Associates, 171
Lee, C. D., 29
Leo, R., 240
Lieblich, Israel, 51, 290
Lie detectors, human, 93–108; see also Poly-
        graph; Polygraphers

Lie detector test questions, 115–119, 143–147
    "control" questions, 9–11, 16, 31,
        116–117, 120–122, 138, 140–142
    irrelevant questions, 115–116
    lie-control questions, 117–119
    relevant questions, 16, 117
    truth-control questions, 18–19, 31, 128,
        143–147
Lie detector tests, See also Clinical Lie Test;
        Control Question Test; Criminal
        investigation, Directed Lie Test;
        Employee screening, Guilty
        Knowledge Test; Positive Control
        Test, Relevant/Irrelevant Test,
        Searching Peak of Tension Test,
        Truth Control Test
    accuracy of: see particular test in question
        the Murray accuracy study, 71–73
    American vs. European attitudes to-
        ward, 51
    assumptions of: see particular test in question
    attachments to the body in, 12
    base rates in, 256–259
    countermeasures against: see particular
        test in question
    critical discussion of: see particular test in
        question
    as evidence in courts, 250–262, 306–307
    example of method, 7–19
    in foreign countries, 51
    numerical scoring of, 33
    opposition to, 46
    polygraphers' claims of accuracy, 64, 69–73
    in psychopathic lying, 267–270
    reasons for taking, 2–3
    reliability: see Reliability of polygraph tests
    science and, 49–52
    summary of, 189–191
    uses of, 5–6
    validity: see Validity of polygraph tests
    voice stress analysis compared with,
        167–171
    why actual lie detection is probably im-
        possible, 62–66
Lindberg kidnapping case, 25. 41
Lippold, O., 166
Lombroso, Cesare, 24
Lonetree, Sgt. C., 245
Long, Russell, 233
Los Angeles Police Department, 295–296

Luria, A. R. 24
Lying
  in animals, 21–22
  base rate of, 256–258
  as bloodless violence, 59
  defined, 22
  guilty feeling and, 5, 60
  ordeals for detection of, 23–24
  physiological indications of, 24, 60–63
  and the psychopath, 267–268
  specific lie response, 60–66
Lykken's Law, 74–75
Lynch, B., 167

Marine guard, scandal, 245–246
McCoy, Judge, 250
McFarlane, R., Col., 218
McQuiston, Charles, 41, 165
McVeigh, T., 299–301
Mackay, Charles, 24
Major C., 214–216
Marañon, G., 61
Marcuse, F.L., 83
Marital problems, 23–24
Marston, William Moulton, 25–27, 28, 64–65
Matte, J.A., 189
Meehl, P.E., 67
Miller, A.L., 45
Miller Analogies Test (MAT), 75
Minor, Paul, 69
Morand, Justice Donald B., 48–49, 102, 106, 136
Moss Committee, House Governmental
    Operations Committee, 163–164
Munsterberg, Hugo, 25
Murrah Federal Building bombing case,
    299–301
Murray, K.E., 71–73

Naval investigative Service (NIS), 245
Nachson, Israel, 169–170
Newberg, D.C., 65
Nixon. Richard, 41, 216
Northwestern University School of Law,
    Scientific Crime Detection labora-
    tory, 29
Numerical scoring of polygraph charts, 33

Oberdorfer, D., 247
Office of Technology Assessment (OTA).
    37, 226

Ordeals, used for lie detection, 23–24
Orne, Martin, 259
O'Sullivan, M., 103
Oswald, Lee Harvey, 41

Parkinson, C.N., Parkinson's Law, 74
Pathometer, 64
Patrick, C., 86, 134
Peak of Tension test, 30, 147–148, 190; see
    also Guilty Knowledge Test
Peer review, 49–51
Personnel screening: see Employee screening
Personnel Security Inventory, 225
Peter, L.J., the Peter Principle, 74
Pillsbury, M and Stinger missiles, 218
Pius XII, Pope, 47
Plants, experiments with, 32–33
Plea bargaining in Manhattan courts, 57
Plethysmograph, 24
Podlesny, J., 290, 305–306
Police force applicants, polygraph screen-
    ing of, 236–237
Polygraph
  history and development of, 21–30
  physiological responses recorded by, 12
  picture of, 9
Polygraph charts
  global scoring of, 91, 218
  independent evaluation of, 85
  numerical scoring of, 32–33
Polygraph Protection Act: see Employee
    Polygraph Protection Act
Polygraph screening tests, 151–161, 190
  assumptions of, 158–160
  test format, 156–158
  validity of, 160–161
Polygraph tests: see Guilty knowledge
    tests; Lie detector tests
Polygraphers
  in clinical lie testing, 93–108
  in employee-theft investigations, 197–212
  as expert witnesses, 260
  "friendly," 259
  judgement of lie-test accuracy, 68–74
  as private detectives, 197–212
  training of, 29, 30–31, 35
Positive Control Test (PCT), 140–143, 190
  assumptions of, 140–143
  validity of, 143
Positron tomography, 62

Posttest interrogation, 92
Premack, Ann and David, 22
Psychological Stress Evaluator (PSE), 165–171
Psychological test, 34
Psychopathic lying, 267–270

Rape, lie-testing victims of, 2, 120, 125, 140–141
Raskin, David C., 35–36, 37, 51, 65–66. 69, 82, 116, 119, 130–133, 179–180, 191, 259, 269–271, 276–277, 305
Rasmussen, D.K., 178–179
Reagan, Ronald, 42, 216
     and the Defense Resources case, 216–217
Reid, John E., 29–32, 68–69
     and Associates, 29
     clinical lie test, 93–108
     Control Question Test, 115
     guilt-complex question, 143
Reid Report, 43, 223–226
Reilly, Peter, case, 240–242
Relevant/Irrelevant test, 109–113, 190
     assumptions of, 110–112
     validity of, 112–113
Reliability of tests, see also Validity of tests
     defined, 76–77
     the Horvath reliability study, 79
Rights
     of employees accused of theft, 208–221
     of employers, 207–208
     of informed consent, 209
Rollison. M., 289
Romig, C. A. 236
Rorschach Inkblot Test, 256
Rothwax, Judge Harold J., 57

Sabotage, 199
Saxe, Leonard, 37
Schachter, J., 60
Schizophrenia, 256
Schwartz, G., 166
Science and the polygraph, 49–52
Scientific opinions about polygraphy, 175–188
     two prior polls, 177–178
     two recent polls, 179–188
     why even scientists are not 100% anti-lie detector, 187–188
Screening tests: see Employee screening

Searching Peak of Tension Test (SPOT), 148–149, 190
Secret Service, U.S., 130–132
Self stimulation to defeat lie detection, 274–277
Sexual abuse, allegations of, 251–253
Simpson, O.J., 296–299
Single-culprit paradigm, 82–83
Sister Terressa, 223, 225
Skinner, B. F., 237
Skolnick, Jerome H., 48
Slowick, S., 129
Smith, G., 166
Smith, R.E., 46
Snyder, LeMoyne, 237
Socialization
     scale, 225
     of children, 227
South Bend Lathe Corporation, 233
Specific lie response, 60–66
Stage fright, 103
Stanton Survey, 224
Steiner, George, 22
Stimulation test (stim test), 14, 19, 99–101, 191
Sucker, synonyms for, 23
Summers, Father Walter G., 64
Susuki, A., 51
Sweat box, 28, 242

Third degree, 242
Tillson, John, 217
Tobin, Y., 166
Torture, confessions obtained by, 23–24, 28–29, 242–243
Tremor in the voice: see Voice stress analyzers
Trials: see Courts
Truth Control Test, 143–147, 190
     assumptions of, 144–146
     example of, 145
Truth Phone, 171
Truth verifier, 83–88

Validity of polygraph tests, see also under specific names of tests
     bad validity studies, 128–132, 135
     the Barland & Raskin validity study, 82
     burden of proof in, 102, 136, 142, 159, 213, 279
     criterion measures in, 85–87

Validity of polygraph test (*cont.*)
   defined, 79–84
   a definitive study of, 87–90
   as evidence in court, 255–260
   methods of estimating, 84–87
     laboratory studies, 84–85
     real life studies, 85, 87
   scientific opinion on, 184–186
   vs. reliability, 76
Voice stress analyzers, 41–43, 163–173
   assumptions of, 165–167
   validity of, 167–173
Vollmer, August, 27

Weir, R.J. Jr., 66
Wicklander, D., 129
Wigmore, John Henry, 57
Williams, Cpl. R., 245
Wilson, G. 216
Wiretaps, 46
Wonder Woman, 25

Yancey, W., 295–296
Younger, I., 65

Zone of comparison test, 190; *see also* Control Question Test